Building *Badger*
& the Benford Sailing Dory Designs

*Anthony Swanston's 37½' **Wild Fox** of Belfast, NI*

by Pete Hill and Jay Benford

Easton, Maryland

Other books by Jay Benford:

Cruising Boats, Sail & Power, 4 editions in 1968, 1969, 1970 & 1971. Design catalog and article reprints. OP*

Practical Ferro-Cement Boatbuilding, with Herman Husen, 3 editions in 1970, 1971 & 1972. Best-selling construction handbook, a how-to on ferro-cement. OP*

Designs & Services, 7 editions, 1971, 1972, 1987, 1988, 1990, 1993 & 1996. Catalog of plans & services of our firm. This material is now all online at www.benford.us

Boatbuilding & Design Forum, 1973. A monthly newsletter with more information on ferro-cement boatbuilding & other boatbuilding information. OP*

The Benford 30, 3 editions in 1975, 1976 & 1977. An exposition on the virtues of this design and general philosophy on choosing a cruising boat. OP*

Cruising Designs, 1975, 1976, 1993, 1996 & 2003. A catalog of plans and services and information about boats and equipment.

Design Development of a 40 meter Sailing Yacht, 1981. A technical paper presented to the Society of Naval Architects & Marine Engineers, at the fifth Chesapeake Sailing Yacht Symposium, and in the bound transactions of that meeting.

Cruising Yachts, 1983. A hard cover book with a selection of Benford designs covered in detail, including several complete sets of plans, a lot of information about the boats and how they came to be. Eight pages of color photos. OP*

The Florida Bay COASTERS, A Family of Small Ships, 1988. A book of study plans of these Benford designed freighter yachts. OP*

Small Craft Plans, 1990, 1991, 1997 & 2005. A book with 15 sets of full plans for 7'-3" to 18'-0" dinghies and tenders.

Small Ships, 5 editions, 1990, 1995, 1995, 1997 & 2002. A book of study plans for Benford designs for tugs, freighters (like the Florida Bay Coasters), ferries, excursion boats, trawler yachts, houseboats & fishing vessels. Eight pages of color photos.

Pocket Cruisers & Tabloid Yachts, Volume 1, 1992 & 1996. A book with 6 complete sets of plans (11 boats including the different versions) for boats from 14' to 25', including 14' & 20' Tug Yachts, 17' & 25' Fantail Steam Launches, a 14' Sloop, a 20' Catboat, and 20' Supply Boat & Cruiser.

Catboats, 2009. A book with six complete catboat designs, from 17' to 22', included.

* OP = Out of print

Copyright © 2011
by Jay R. Benford
ISBN 10: 1-888671-28-9
ISBN 13: 978-1-888671-28-5

All rights reserved. No part of this publication may be reproduced, stored in a retrieval system or transmitted in any form or by any means, electronic, mechanical, photocopying, recording or otherwise, without prior permission in writing from Tiller Publishing.

Graphic design & production:
Scribe, Inc.
842 S. 2nd St.
Philadelphia, PA 19147

Printed in the USA by:
Victor Graphics, Inc
1211 Bernard Drive
Baltimore, MD 21223

Questions regarding the content of this book should be addressed to:

Tiller Publishing
29663 Tallulah Lane
Easton, MD 21601
410-745-3750

Dedication:

To the builders and owners of our dory designs, who kindly shared their photos with us.

And, of course, Dona. . . .

Table of Contents

Notes:

For this book, we decided to put the designs in chronological order, rather than by size, so that you can follow the development of the designs and how we've evolved what we have created over more than four decades. Some of the changes were driven by what the clients requested and some came about as our experience and thinking grew.

Our drawing numbering system for our designs started with the design I numbered one. I've known some designers who started numbering at 60 or 100 or some other number, not wanting the folks seeing the number to think they were just beginning their design career. Perhaps I was oblivious to this nuance, or it didn't matter to me. Whatever the reason, it was a lot of decades ago—I was still in school at the time—and that's just the choice I made. . . .

Some of the drawing pages are cut-and-paste assemblies of several drawings to fit the page sizes of the book. There are some details that are not included here, but are part of the actual building plan sets. We have striven to reproduce all these plans accurately so that they can be scaled with an architect's scale. Most will be found to be at 1/8" = 1'-0" except the 45' dory drawings, which will be 3/32" = 1'-0". The 32' St. Pierre and 23' Knockabout are at 3/16" = 1'-0". Some of the details were that done at larger scales originally will be here at larger scales than 1/8" = 1'-0".

Page	Design Number	Size & Name
4		Introduction
7		Building **Badger** by Pete Hill
70		The Most Economical Offshore Cruisers
74	17	22' Sailing Dory
79	21	26' Sailing Dory
84	24	27' Sailing Dory
85	27	26' St. Pierre
89	29	23' St. Pierre
91	30	19' St. Pierre
94	32	30' Sailing Dory
102	33	27' D. E. Ketch & Schooner
104	36	32' Sailing Dory
108	41	26' V-Bottom Ketch
111	113	32' St. Pierre
116	127	36' Sailing Dory **Donna**
129	140	60' Aluminum Sharpie
131	170	34' Sailing Dory **Badger**
146	174	37½' Sailing Dory & Trawler
158	179	23' Knockabout **Sourdough**
165	274	26' Twin Keel Cutter
171	301	36' Power Dory **Proctor**
176	323	45' Sailing Dory
180	370	30'-5" & 31'-8" **Baby Badger**
187	384	45' Sailing Dory
191		Glossary

Below—20' variation idea based on 19' St. Pierre Dory.

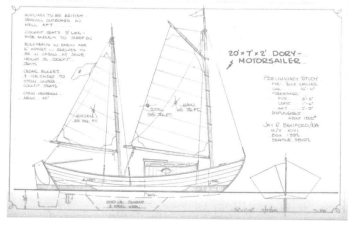

Introduction

By Jay Benford

When my folks realized that launching me was forthcoming, they sold their Snipe and got a 21½' cruising boat, that had been built by a local Dutchman for his own use. Built originally as ***Jericho***, they renamed her ***Flying Dutchman***. My folks took me sailing before I could walk and I became an active part of the crew as I got old enough to be useful. I went along with my father when I was 12 and he was taking the US Power Squadron introductory course. I passed the final test and got a letter saying I would be welcome to join the Squadron when I came of age.

In the meantime, I was haunting the public library and reading all the materials on yacht design and boatbuilding I could find. My folks were persuaded to let me loose with the boat the summer when I was 15. A friend and I sailed it from Rochester to Fair Haven, 45 miles to the East on Lake Ontario. On the way, the engine threw a rod and we ended up sailing the rest of the way, on a day with light conditions. While we waited for the arrival of parts, I ended up

Flying Dutchman *the summer before I lived aboard.*

living on it for a month or more there—and got a good deal of practice in making landings under sail. In retrospect, it was quite brave of my folks to turn me loose like this, but the boat and I survived quite well.

Along with me on this cruise were perhaps a dozen of the Motor Boating Ideal Series books. These were each full of dozens of designs, mostly by William and John Atkin. I loved their approach to creating a wide variety of small cruising boat designs. By careful study of the drawings and descriptions, I was able to learn a lot of boatbuilding terminology which served me well later on. Amongst the Atkin designs were some based on dory hull forms. Several years later, when I was fortunate enough to apprentice with John Atkin, we did a series of dory designs for Captain Jim Orrell's Texas Dory Boat Plans.

Later, upon setting up my own self-employment office in Seattle, I wrote to Captain Orrell and inquired if some of my designs might be useful for his plan sales business. This led to doing the 22' and 26' Sailing Dory designs and the 19', 23', and 26' St. Pierre Dory plans.

For myself, I was dreaming of a cruising boat for myself for Puget Sound cruising. Not having deep pockets, I recalled the appeal of the quickly built dory designs. So, I did designs for 30' and 32' Sailing Dories and got estimates on having them built. The 30' and 32' were included in my plans catalog and stirred a bit of interest. An inquiry that I got led to my creating the 26' V-Bottom Ketch, a design that looked much like the 26' Sailing Dory, using her profile and rig.

*Christening of **Donna** in Vancouver, BC.*

Then we got a client for a larger version of the dories. This became the 36' Sailing Dory ***Donna***. She was built by a shoemaker who'd never built a boat before, but possessed a very strong work ethic. Building in Alberta, where there weren't other boatbuilders locally to visit and talk to on how to handle different construction tasks, he spent just over a year on her. He had her trucked to Vancouver and invited us to join him for the launching, which we did.

The builder/owner/skipper went on to do a lot of ocean voyaging on her. After several ocean crossings, he came back to us and asked for a larger sail plan, saying he wanted to have a taller rig to catch the wind as he was surfing in the Pacific. The resulting taller Marconi ketch rig did what he wanted and has offered an extra choice to other builders, in addition to the junk schooner and cutter rigs.

We had another request for a dory a little bigger than the 30 and smaller than the 36, so we created the 34. The original version had a trunk cabin and a cutter rig. We also created a raised flush deck version and the Hills in building ***Badger*** did the first junk rig for her. They used Hasler and McLeod's **Practical Junk Rig** book for a guide and did a good job. In our visits with them, they shared the marked up prints they'd used and we had a starting point in making the junk rig drawings for her that have been so popular. We also created a gaff cutter rig which several builders wanted to use on her.

The 37½' Sailing Dory was done for someone who wanted a little bigger version of ***Badger*** and we've since done a variety of rigs for her too. Her design was done by tilting the stemhead forward a bit and slightly extending the sternpost to allow for the bulwarks at the stern. The structure and underbody are virtually identical with the 36 and many of the plans are shared between the two versions.

A concept for a three-masted junk rigged schooner, with a modified dory-like hull was done, but it never got past the very preliminary design stage.

We did do two 45' dory designs. The first was a sort of motorsailer and we looked at a variety of rigs for her. The client's theory was to have several staterooms that would make her suitable for charter work. The second one is a larger junk schooner that has a lot of room for living aboard and cruising. She has a rather thick fin keel which houses the mechanical-electric propulsion besides the ballast.

The most recent addition to the dory range is the trawler version of the 37½' and this is shown in the chapter with the sailing version. It's a very different approach to doing long distance voyaging under power.

Check out all these designs in detail in the chapters of this book! . . .

Above—Pete Hill in the designer's office in 1991.

Pete Hill's fine workmanship is evident in these photos of his **Badger**. They were taken by her designer on a visit aboard (circa 1992) while **Badger** was in Florida.

Building *Badger*
by Pete Hill

Introduction

Annie and I built **Badger** because we had specific ideas about the sort of cruising boat that we wanted and there was nothing that fitted our ideas available on the second-hand market, at least not at a price we could afford. We were also attracted to the idea of building a boat for ourselves, for its own sake.

Make no mistake, building a boat is a major undertaking that may well take years of your life, and a great deal of your money. Be realistic about yourself and your circumstances. Do those around you support your enthusiasm? Have you got the drive to sustain yourself through the hard labour and the low times when things aren't going well? The satisfaction of sailing a vessel built with your own hands is countered by the depression of walking away from an unfinished project. Think long and hard; don't rush into building.

If you do decide to build, or if you are still considering whether building is for you, I hope the following advice will be helpful. Advice is always fun to give—you get to tell people what to do, without the responsibility of living with the results.

Money

It probably won't come as any surprise to you that you're going to need money to build your boat. One of the advantages of building over buying is that you don't need all the money at once, but you do need a regular supply of it. Ideally, you will have enough money in the bank to buy all the materials, plus a reserve because it's going to cost more than you think. Amateur boatbuilders tend to be optimists, otherwise, they would never start, and they tend to underestimate how much it will cost and how long it will take. If you don't have all the money up front, you will probably have to go out and earn it, which, while socially acceptable, does cut into the time spent building. Remember also that time equals money. For example, most of the solid timber we used was second-hand, which gave us good quality timber at a very reasonable price, but it took a lot of time sawing it up. Making our own sails saved a lot of money, but they too, took many hours to make.

Building site

Unless you are lucky, finding somewhere to build is a big problem, especially in an overcrowded country such as England. A large, heated shed with electricity and water, situated right next to where you live is the ideal, but one that is rarely achieved. If you are thinking of building next to your house, make sure that *all* your neighbours are happy about it and that they will remain happy, until the boat is completed. They probably didn't move into their house to have a large polythene shed next door, with a whining power plane to listen to at all hours. Even if the neighbours are happy (in which case, they probably don't know what's involved), check with the local council about planning permission.

Travelling time to and from the building site is an important consideration. Apart from the time and money wasted by travel, it may mean that it isn't worth using those few hours that you have spare each evening, which all add up.

If you manage to rent a site or a building, make sure that you give the owner a generous estimate of the time that the project will take.

A building site without electricity is a major disadvantage, but you can get over the problem with a generator, in which case a diesel one will probably pay for itself. Once again, however, consider the fact that the noise of the generator will constitute a nuisance to many people.

If you need to build a shed, the Gougeon brothers' book or Buehler's Backyard Boatbuilding, will tell you how. (See section 1 below)

Time

How long will it take to build the boat? This is a tricky question. Annie and I spent three years building *Badger* and between us, we probably put in at least 10,000 hours. I'm sure that a professional boatbuilder could do the job in less than 5,000 hours. The extra hours that we took were due to several reasons. Lack of skill meant that it took longer to do the work than it would take a professional—this disadvantage was reduced with time as our levels of skill increased. Lack of access to professional quality power tools was another reason—these can save a lot of time. As I mentioned earlier, doing things in a way that saves money also takes a lot of time. A professional has his hourly rate and all money saving is based on that figure. On *Badger*, we also spent a considerable number of hours in working for a good finish, such as putting the veneers on the bulkheads, insulating the hull and laying teak decks. If you are happy with a much more basic finish, you can cut out a lot of time by simply having painted plywood on deck and below. On the other hand, don't fall into the trap of saying "Let's go sailing now and we'll finish it off as we go along." Chances are that you won't; you'll live with it the way it was launched for years.

If you build your boat to a similar standard to *Badger* and try to do it as cheaply as possible, our 10,000+ hours is probably realistic. Think about this. If there is only one of you building and you only work at the weekends, say twenty hours a week, that's only 1,000 hours a year, which means that she will take you ten years to build! This is why Annie and I put in all the hours we could and went sailing after three years. Life is too short to spend eight or ten years building a boat, if the object of the exercise is to go sailing.

Skills

To someone who has never built a boat before, the prospect is daunting. "Do I have the skills to carry out such complicated work?" might well be the average response. Some degree of skill is certainly required, but the average 'handy' person can soon acquire this. If you have never done any woodwork, it would be prudent to go to evening classes to learn how to use tools. The joints used in building a plywood boat are straightforward and if you can cut to a line and plane a piece of wood to shape, then you will probably cope. If you have doubts about your abilities, try building a dinghy first—the skills needed are the same, it is merely the size of the project that differs. Ideally, the dinghy can be used as the tender to the yacht that you are planning and that will be one less job to do before you go cruising.

The great advantage of a design like *Badger* is that the construction of the hull is straightforward. You can't get much simpler than a dory, built out of plywood. By the time you have finished the hull, you will have learned a lot and will be more confident about tackling the rest of the construction.

Management

To build a boat for yourself is the equivalent of setting up a small boatbuilding business. The fact that this particular business in non-profit making does not alter the fact that it still needs managing.

You will need to work out a budget for the project, make an estimate of the materials needed and the cost of overheads (site rental, electricity, tools, travel costs, etc) and ascertain whether this matches the money available for the project, or your expected surplus income. Materials will have to be ordered so that they are on hand as required; you will be doing the same job as a buyer for a boatyard. Find out how long the lead time is on ordering the various materials and get the best price you can. This usually means buying in bulk, if you have the money available. An offer to pay in cash can also produce a discount.

If more than one person is involved in the project, more management is required. You need to organise who does what so that too much time is not wasted waiting for someone else to finish a job so that the next person can get on with their bit. Sharing the creative work, as well as the tedious jobs, keeps everyone motivated—ensure that you are an equal opportunities employer.

In order to progress smoothly, the project as a whole will need managing. Plan the work ahead, so that everything comes together. Think through what you want to do next and try to see if there are going to be any hold-ups, such as waiting for glue to harden. You can then organise your work more efficiently. Long term planning is also required; this is especially important if the weather plays a rôle in what you can do. Having the materials on hand as they are required is an important aspect of the planning. Keep a check on your progress and, if the whole project seems to be taking too long, analyse what the problem is. You probably started building the boat in order to go cruising; make sure that you keep this goal insight. Don't get bogged down in building the perfect boat that would make a grand piano look shoddy by comparison.

Power tools

Power tools are not essential to build a boat, and if you have no electricity available, you can do without them. However, they do save a great deal of time and often produce much better results for the essentially unskilled handyman. Whatever your budget, buy the best quality tools, designed for the professional. The handyman quality is generally a waste of money, in the long run. You are building a relatively large boat, not putting up kitchen shelves, once every few years. The professional quality tools will give better results, last the life of the project and you will probably be able to sell them at the end.

How many tools you have depends on your budget. The obvious ones to start with are a power plane, a jigsaw, circular saw, table saw, drill and sander. If you have the money a power mitre saw and a router are both very useful, especially when fitting out the interior. Tungsten tipped blades and cutters are always worth the extra money. They do need to be professionally sharpened, but they stay sharp for a long time. If you are going to be dealing with a lot of second-hand timber, have a spare, cheaper blade for the initial cut, when you might go through a nail that you missed when examining the wood.

Plywood

There is a huge difference in the costs of different types of plywood. Top grade, marine plywood such as Bruynzeel™ or Aquaply™ is very expensive, but of excellent quality. If your budget doesn't stretch to that, then buying cheaper plywood takes a lot more care. We used WPB (will not delaminate, even if boiled in water), Lauan (Philippine mahogany), exterior-grade plywood to build *Badger*. It was of very good quality, with no voids in the veneers and the offcut that we still use as a board for gutting fish is showing no signs of delamination or rot, despite living in a deck locker for the past fifteen years. I have seen several so-called 'marine' plywoods, with voids in the veneers, that cost two to three times as much as our Lauan. Hardwood veneers will make it a lot easier to get a good finish.

The two things that are essential in plywood are that the glue should be waterproof and that it should be free from voids, which is where the rot starts. Never buy plywood unless you have inspected the quality first, or know the brand name. Building with epoxy, to seal the end grain, gives plywood a long life and after fifteen years of sailing Badger, we have found no problems with our exterior grade plywood. You pays your money and you takes your choice. *Buehler's Backyard Boatbuilding* discusses the choice of plywood more fully.

Health

Don't underestimate the health hazards of boatbuilding. Hand tools can cut you badly if not handled properly. Power tools can maim you for life. Sawdust is an irritant, so wear a dust mask. When chips are flying, wear safety glasses. Probably, the best eye protection is a visor, as used for chain sawing; glasses can steam up.

The most dangerous substance that you will be using is epoxy resin. The hardener is extremely toxic and it is very easy to get sensitised to it. What this means in practice, is that you come out in a very painful rash (contact dermatitis), if you are exposed to uncured epoxy on a sufficient number of occasions. To control this, you will need to use steroid creams on a regular basis: not a happy thought. You cannot be too careful about keeping epoxy off your skin. Use a barrier cream on all exposed skin, wear rubber gloves and use overalls, which are frequently washed. Work as cleanly as possible. If you become sensitised, it lasts for life and you may also find that polyester resin and polyurethane paints will bring you out in a rash. Once sensitised, the fumes of the hardener will be enough to start the rash off again. It is a fact that many people who build a boat, using epoxy resin, become sensitised—I am one of them. If you are scrupulously careful, you probably will not.

Books

I am a great believer in learning from books—read as much as possible about boatbuilding before you start. A book that I would certainly recommend is:

The Gougeon Brothers on Boat Construction. This book is essential reading if you are going to use epoxy. I found it an excellent general boatbuilding book and all the advice in it is based on practical experience.

What now follows is how we built *Badger*. I'm not saying that this is the way to build a boat like *Badger*, just that it's the way that we did it. I hope it will give you an idea of what is involved. You may well come up with better ways of doing things, but if not, at least here is a method that worked, got the boat built and well enough that we are out here, still sailing her fifteen years and 100,000 miles later.

Building *Badger*

1   Having found and rented a site, we set about building a shed. The frames were two concentric 1/2" [12 mm] plywood hoops, spaced by softwood 2" [50 mm] x 4" [100 mm] blocks. They were made up on the ground, by screwing the blocks at regular intervals to one length of plywood. This strip was then bent to shape on edge, and held there by bricks while the second length of plywood was screwed to the other side of the blocks. The resulting frame was light and reasonably rigid.

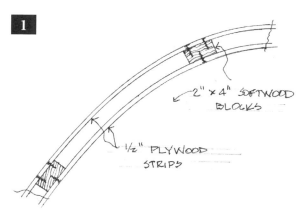

2   The frames were erected about 4ft [120 cm] apart and held together with longitudinal timbers and diagonal bracing.

3   A thick black (ultraviolet-proof) polythene sheet was pulled over the top and nailed in place around the edges.

4   Clear polythene sheet was used to cover the end walls and along one side of the shed, to let light in. The shed was a bit flimsy, but lasted long enough to build the hull. A bad winter gale demolished it.

5   Lofting the hull, full size, was the first job. A framework of second-hand 2" [50 mm] x 4" [100 mm] timber, was laid on the ground and covered with sheets of 3/8" [9 mm] plywood (which were later used in the hull construction). A long length of clear 1" [25 mm] x 1" [25 mm] timber was scarfed together for the fairing batten.

6   Lofting a dory hull was very straightforward, because of the simple, hard-chine shape. Only the sheer and chine needed to be faired, in order to check the offsets. The shape of each bulkhead is taken from the lofting table, as is the profile of the curved stem.

7   The bulkheads were cut out of plywood. Most of the bulkheads could not be made from one piece of plywood and so were joined using a half lap. The waterline was marked on the edge of each side of the bulkhead.

8   To cover up the joints and to provide an attractive finish, the bulkheads were covered with 1/8" [3 mm] veneers of pitchpine, glued on with temporary staples. We cut the veneers from second-hand wood on a circular table saw, but had no thickness planer, so the veneers varied in thickness.

9   When the glue had set, the staples were removed and the surface planed smooth, using a hand power planer, set to a fine cut. The surface was then sanded smooth.

10 Using a router and a straightedge, a veed groove was made at the edge of each veneer, to give the effect of tongue and groove planking. The straightedge was clamped at the top and bottom of the bulkhead, half the width of the router base from the joint in the veneers. The router in the picture is, in fact, a power drill with a router attachment—we couldn't afford a router at the time.

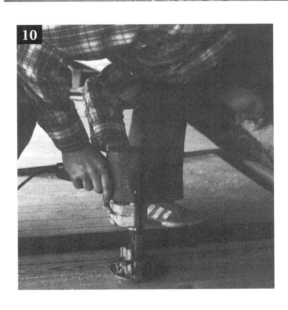

11 2" (50 mm) x 1" [25 mm] framing was glued to the edge of the bulkheads, remembering to put it on the side that would be planed off to take the curve of the hull. The notches for the chine log and sheer clamp were cut and the faces of the bulkhead coated with clear epoxy, to protect them during the building process. The midships bulkhead here, is in two halves and is joined together with one of the heavy floor timbers.

12 The stem was laminated over a simple strongback, which was made from scrap timber. The shape of the strongback was taken from the lofted profile of the stem.

13 The bow knee was also laminated, shaped and glued to the stem piece.

14 A mast step was built into the forward knee, with heavy plywood sides and a drain hole in each side, at the bottom of the step.

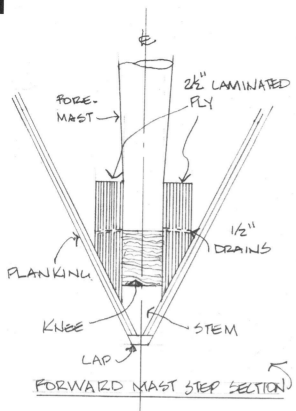
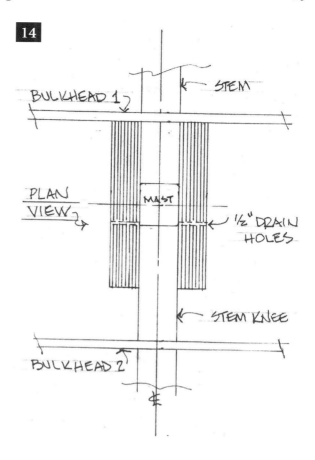

14

15 The forward side of the step was used to locate the forward bulkhead, and wooden chocks on each side of the knee did the same for No 2 bulkhead.

The bulkheads were glued in place.

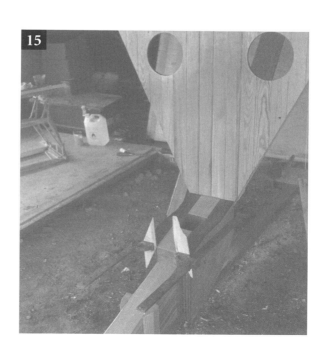

16 The sternpost is straight and was laminated up from several thinner pieces. The stern knee was glued on in the same manner as the forward knee. Bulkhead No 6, at the stern, was made from 2" (50 mm) x 1" [25 mm] timber and joined at each corner with wooden gussets. This bulkhead was joined to the stern piece with a notch cut into the knee.

17 The next stage was setting the stem, stern and all the bulkheads in place. The stem is the highest piece on the hull and had to be set up first, as low as possible to make it easier to work on the hull, which was to be built upside down.

18 Once the bow section was set up level, the next bulkhead could be accurately positioned. We made a stand for each bulkhead instead of constructing a strongback. Using a plumb bob, the centreline of the bulkhead was lined up with a centreline wire, stretched just above the ground. The plumb bob was also used to check that the bulkheads were vertical. A water gauge (a length of clear tubing, almost filled with water) was employed to get the height of the bulkhead correct, using the waterline marked on the edge of each side of the bulkheads. The bulkhead could be moved up or down in the stand and clamped in place. It was essential to check that the distance between the bulkheads was correct and that each bulkhead was at right angles to the centreline wire. This was checked by measuring the diagonal distances between the bulkheads. Doing this was very fiddly as each time one bulkhead was slightly adjusted, all the measurements had to be checked again. However, it is one of the most important jobs, because the final shape of the hull is determined here, and it is worth being patient and careful.

19   Once all the bulkheads and the stern were erected, we double checked all the measurements again. The next task was to tie all the bulkheads together, with temporary battens nailed on.

Using a faring batten, the notches at the chine were bevelled to follow the curve of the chine. The chine log was laminated in three pieces and the timber was cut to the correct width and bevel on the saw table. The first layer was scarfed to length, fitted at the bow and glued in place at each bulkhead, using temporary fastenings of steel screws, which were first greased, for ease of removal. Fitting the chine at the stern was tricky, because it had to be cut to the correct length and angle. In the picture, note that the mast step and the stem have not yet been faired in.

20   The chine log was built up, putting the laminates on each side alternately, in order to keep the pressure even. The second and third layers were glued and scarfed to length in place. Again, we used greased, steel screws as temporary fasteners until the glue had set. Once all the layers were fitted, the whole bulkhead assembly became very rigid.

21   After bevelling the sheer clamp notches, we tried fitting the first layer in one 5" [125 mm] wide piece. Because of the sheer of the hull, there is a lot of edge set and this caused the plank to twist out of true between each bulkhead.

22 We got around this problem by sawing the plank into three and, in effect strip planking the first layer. Thin dowelling was used to keep the strips in line with each other.

23 The second and subsequent layers were glued on and scarfed to length using the full width, with no further problems.

24 With the sheer clamp laminated up, the basic framework of the hull was complete. The inside edges of the chine log and sheer clamp were sanded smooth. It was much easier to do this now, before the planking went on.

25 The bottom of the boat was faired up ready for the first layer of planking. With the power plane, the chine logs, stem and stern were planed smooth and faired from side to side, using a batten to show the high spots.

26 The floors are the heavy timbers, which span the bottom of the hull and through which the bolts for the ballast keel pass. They were laminated up to size, epoxy coated and sawn to fit between the chine logs.

27 The floors were glued in place, held by clamps. Once the glue had set, a 3/4" [18 mm] dowel was glued through both floor and chine log at an angle, to tie them together. Note that the chine log was not notched to take the floor. The floors were faired smooth, with the rest of the bottom.

28 In preparation for planking the hull, sheets of plywood were flow-coated with epoxy. These formed the inside layer. Flow-coating effectively puts the equivalent of three coats of epoxy on the surface in one go, with no sanding between coats. Not only does this save a lot of time, but it also gives a good, smooth finish.

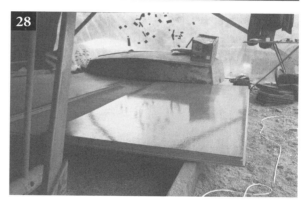

29 Using a scarfer attachment on the portable circular saw, the plywood was scarfed, where necessary, to the required width at bow and stern on the bottom of the boat. The scarfer gives an 8:1 bevel to the plywood.

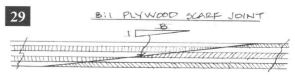

30 The bottom of the boat is built of three layers of 3/8" [9 mm] plywood and the topsides are of two layers of 3/8" [9 mm] plywood. The first layer of plywood was put on the bottom. The plywood was scarfed at the joints, which were arranged to be over bulkheads or floors.

31 We decided to stagger the layers of plywood on the hull, so that as much of the endgrain of the plywood as possible, would be covered up, once the first layer was put on the bottom. The next job, therefore, was to put the first layer of topside plywood on. The sheer clamp, chine log, stem, stern, bulkheads and the edge of the plywood on the bottom of the hull, were all planed fair

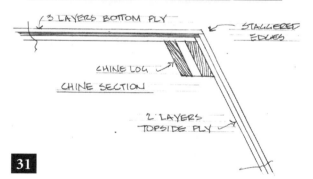

32 A fairing batten was used to ensure that the plywood would lie flat.

33 The bow and mast step were also planed down to fair in with the hull, ready for the first layer of plywood.

34 We decided to fit two stringers to each side of the hull in order to make planking easier and to add additional strength. Notches were cut into the bulkheads to take these.

35 The photograph shows the stringer being fitted at the stern. The vertical piece of wood keeps the stringers in line with the chine and sheer, while fitting them to the stern. This was needed because of the large curve of the stern section.

36 The stringers were glued to the bulkhead using temporary steel screws and permanent wooden dowels.

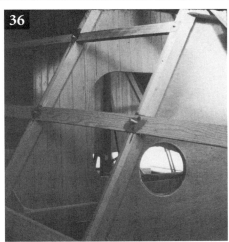

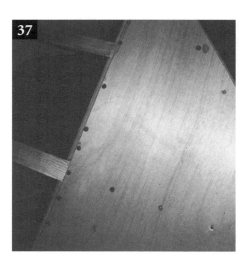

37 Starting at the bow, the first layer of plywood was fitted. Note the temporary greased, steel screws, with washers underneath. The first layer of plywood was butt jointed onto a bulkhead or stringer, wherever possible.

38 In this photograph, the first layer of plywood is being fitted at the stern, the flow-coated inside face can be clearly seen. The coated surface made cleaning up excess glue around the stringers and bulkheads much easier. It was very worthwhile to clean up the glue before it set.

39 The first layer on the bottom of the hull was prepared for the next layer. The edge of the topsides plywood was planed flush and the joints on the bottom were sanded smooth, after which the screw holes were all filled.

40 The second layer was fitted to the bottom. The panels were scarfed together *in situ*, with the joints staggered so that they were in different places from the first layer. The sheets were held with a regular pattern of temporary 3/4" [18 mm] steel screws and by using a washer, we ensured that they did not penetrate the inner layer. Longer screws were used to secure the plywood to the bulkheads, floor and chine log. These were also removed once the glue had set.

41 The second layer was fitted to the topsides. Each piece was scarfed to the next, *in situ* and the joints staggered from the first layer. As on the bottom, temporary 3/4" [18 mm], greased steel screws with washers, were used to hold the plywood in place, with larger ones in the bulkheads, stringers, etc.

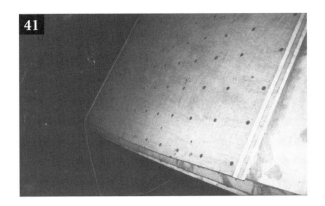

42 Before fitting the final layer to the bottom, large dowels were glued into the stem and stern knees. These took the place of metal bolts, specified in the plans.

43 Once the final layer of plywood was on the bottom of the hull, the planking at the bow was planed smooth to take the capping piece of pitchpine, which covers the end grain of the plywood. The stern was treated similarly.

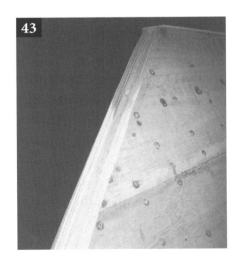

44 The hull was now planked up. The next job was to fill all the holes and joints and smooth them ready for covering with glass and epoxy. We started building the shed in May and finished planking up just before Christmas. Over the New Year, a severe gale demolished the shed, so we decided to leave sheathing the hull until later, but it would have been much easier to have done it at this stage.

45 Having postponed glassing the hull, we prepared it for turning over. Two frames were built around the hull, designed both to protect its surface and to support it at deck level.

46 The hull was turned right way up using ropes and a gang of people recruited from the pub next door. Ropes to either side of the framework controlled the speed at which the hull rolled over. It only took half an hour, but this was as long as such a large crowd could be controlled.

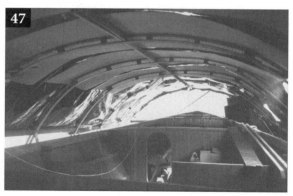

47 Once the hull was the right way up, a new shed, similar to the first one, was built around it. The hull was carefully levelled, so that we could use a spirit level during the rest of the construction.

48 The openings through the bulkheads at either end of the main cabin, were cut out.

49 The forward cabin was fitted out before the deck was put on, which made things much easier. The foredeck framing was fitted, ready for the plywood deck.

50 The completed forward cabin, looking forward. The top lockers on either side hold clothes and the lower shelves hold paperback books. The green tube to starboard contains the sail for the dinghy and the one to port is the sun awning to go over the cockpit. Underneath the shelf, alongside the bunk, is a locker for shoes and linen.

51 Before the foredeck was fitted, the bow section was completed. Left-over epoxy had been poured into the bottom in order to bring the level well above the waterline, so that it could be made into a self-draining chain locker. A hole was made through the hull near the top of the epoxy and a teak finger ring was glued into the hole, to seal the endgrain of the plywood. Slightly thickened epoxy was poured into the bottom until it reached the level of the hole. Note the eyebolt glued in for the bitter end of the anchor cable.

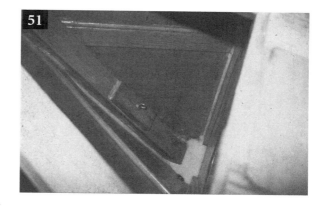

52 A 12" [300 mm] wide piece of 3/4" [18 mm] plywood was set in between the bow and No 1 bulkhead. This was to give extra strength to the foredeck, where the anchor windlass was to be bolted. The whole area between No 1 and No 2 bulkhead was beefed up to take the loads of the foremast. A laminated piece of Douglas fir, 4" [100 mm] thick and 18" [450 mm] wide, was set between the bulkheads. A piece of 1/2" [12 mm] plywood was rebated in on either side of it.

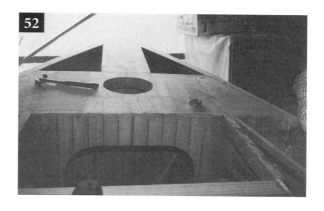

53 The first sheet of plywood was put on the foredeck. The plywood was butt-jointed on the fore and aft stringers.

54 The second layer of plywood was fitted to the foredeck. The joints were staggered from those of the first layer and scarfed, using the same procedures as when building the hull.

55 The interior—locker bottoms, topsides and deckhead—were clad with 1/2" [12 mm] cork, with a thin wood veneer glued on top. This was to insulate the boat and help prevent condensation. Although this has worked well, fire-retardant polystyrene foam might be a better alternative, being both lighter and easier to work.

56 The lazarette was fitted out as a storage area, again before the deck was fitted. The lower lockers were designed to hold several 4 1/2 Imp. gallon [20l] drums, for storing paraffin.

57 The cockpit footwell was constructed, before installing it in the boat. The bottom has strips of teak glued on top and the fours sides were filleted together at the corners and covered with glass cloth and epoxy.

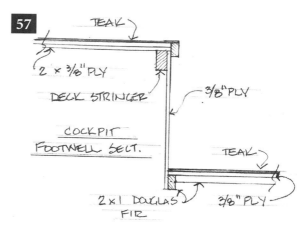

58 The footwell was glued in place between the deck stringer and the after frame, using a spirit level to ensure that it drained aft, where drainage holes were fitted in each corner.

59 The two layers of plywood were fitted to the after deck.

60 A former was made to laminate up the deck beams. The beams were of five layers of 3/8" [9 mm] pitchpine, 1 1/2" [37 mm] wide.

61 The front of the cabin was constructed of 3/4" [18 mm] plywood, with a half lap where it was joined to get sufficient width. Laminated beams were glued to the inside of the plywood. Note the two 1/2" [12 mm] plywood blocks for the forward scuttles. The rest of the plywood was covered with 1/2" [12 mm] cork insulation.

62 The cabin front was dry fitted before finishing.

63 With the cabin front and back dry fitted, the sheer clamp could be planed down for the bulwarks/built-up cabin sides. Using a power plane, the sheer clamp had to be shaped from the flare at the bow and stern to the tumble home of the cabin sides.

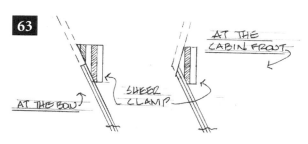

64 Once the fore and after deck edges were planed true, the cabin ends were removed. The teak overlay decks were laid on the foredeck. It was much easier to do this before fitting the bulwarks. In the photograph, a teak plinth is being glued on around the hole for the forehatch. The aluminium hatch was fastened down at a later date.

65 The covering board was fitted and glued into place. The teak used was 3/16" [5 mm] thick in accordance with the method described by the Gougeon Brothers.

66 Using the edge of the deck as a guide, the covering board was routed to an even width, following the curve of the deck edge.

67 The rest of the deck was laid straight. It was dry fitted and then glued down, using temporary greased, steel screws and washers, until the glue had set. The excess glue squeezed up between the planks to form the 'caulking' seam.

68 The screws were removed and the holes drilled out to take a plug. The plugs were tapped in after filling the screw hole with glue.

69 The afterdeck was laid in the same fashion.

70 This photograph shows the fore and after-decks ready for finishing.

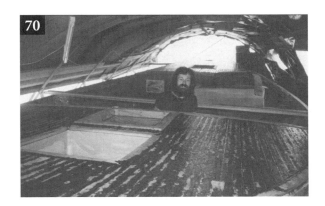

71 The deck was completed using a disc sander to remove the bulk of the glue and any high spots, before smoothing over with a belt sander. The final finishing was done using an orbital sander.

72 The front of the cabin was finished with teak veneers to give a tongue and groove effect—this was done the same way as on the bulkheads.

73 The inside of the cabin was insulated with cork and finished with matt white, melamine sheet.

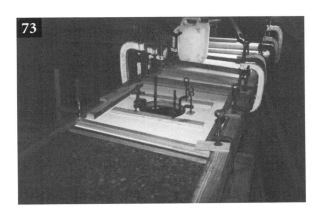

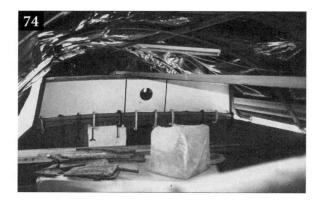

74 The ends of the cabin were glued into place and held with clamps. When the glue had set, 1/2" [12 mm] dowels were glued into the beams, vertically from the bottom. These took the place of bolts.

75 With the front and back of the cabin fitted, work could start on the framework for the raised cabin sides and the centre deck. Two small fore and aft partitions were fitted either side of the companionway. A pattern was made from scrap plywood (a glue gun works well at spot gluing the pattern together). In the photograph, the cut-out plywood is getting a trial fit.

76 A former was made up for laminating the hanging knees that tie the cabin sides and centre deck together. The knees were made larger than shown on the plans in order to help support the extra load imposed on the centre deck by the unstayed mainmast. The former was made by bolting wood blocks to a piece of thick scrap plywood. The laminates were sawn just thin enough to bend around it, about 3/16" [5 mm] and finished up 2 1/2" [62 mm] thick by 1 1/2" [38 mm] wide.

77 A partial-bulkhead, forming the heads compartment was marked out on plywood from a pattern.

78 The main deck support is formed by two longitudinal stringers, either side of the centreline and from three deck beams, supported by the hanging knees and the heads bulkhead. The beams were laminated up on the same former as that used for the cabin ends. The stringers are in place, in the photograph, with the two after partitions and heads bulkhead. The deck beams are lined up and the stringer has been marked for notching.

79 The deck beams and hanging knees were glued into place, together with the heads bulkhead.

80 The raised-deck sheer clamp was then laminated to the cabin ends, the knees and the heads bulkhead. This was very similar, in method, to fitting and laminating the chine log.

81 The raised-deck framework was now complete and ready for fitting the cabin side. Note the extra-thick knees which take the main loadings from the mast. A finishing piece of wood has been glued between the knees on the sheer clamp, to cover the 'strip-planked' piece.

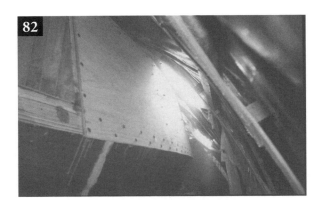

82 It would have been more convenient to have fitted out the centre cabin before the plywood sides and deck went on, but as we were about to be evicted from our building site, the plywood went on to make *Badger* weatherproof. The first layer of the cabin sides is shown fitted in this photograph.

83 The bulwarks at each end were scarfed to length and the insides coated with epoxy, ready for painting, before being glued on. The cabin sides were butted together on the cabin ends and centre knees. The next layer was scarfed in place, with the joints staggered as usual.

84 Before the centre deck was fitted, two layers of 1/2" [12 mm] plywood were rebated between the deck beams, all the way across the cabin, in way of the mast and glued in place. These were to strengthen the deck for the unstayed rig. The deck, two layers of 3/8" [9 mm] plywood, was then laid in the same manner as the fore and after decks.

85 The deck was now on. A 1/2" [12 mm] plywood filler block was put where the scuttles would be bolted on, then 1/2" [12 mm] cork insulation was stuck down and the sides finished off with melamine sheet. Note the plywood scuttle filler extends the full height of the cabin sides alongside the mainmast.

86 We now had to move *Badger* round to a new site. The shed came down and we saw her for the first time with the basic structure completed.

87 We were lent a trailer and, after jacking up the boat, the trailer was run underneath. A friendly local farmer towed us around to our new site, a marina, one mile away.

88 Once off the trailer, in her new home, *Badger* was levelled up. We did not bother rebuilding the shed and just put the old cover over her.

89 Work now started on fitting out the centre cabin. In the galley, the cooker, a Dickenson 'Bristol' diesel range, was fitted athwartships and a shelf was built below for pan stowage, with a storage locker underneath.

90 The rest of the galley was fitted out. The two bulkheads, of 1/4" [6 mm] plywood, which support the drawers, were filleted to the hull. The drawers run on rails of 1" [25 mm] x 1" [25 mm] timber, glued to the bulkheads.

91 The shelves in the lockers were made of slatted wood, to help air circulation. Box fiddles then divided up the space, to prevent the contents from sliding around at sea. The sides of the hull had already been insulated with cork and covered with veneer.

92 The drawers were simply made from plywood, the sides being 3/8" [9 mm] and the bottom of 1/4" [6 mm]. A wedge was glued on the bottom at the front of the drawer to lock it in place when the drawer is shut, and a stop was fitted near the back, to prevent the drawer from coming out.

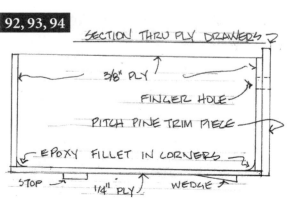

93 The pieces were butt joined and glued, then all the corners were filleted.

94 A trim piece of pitchpine was glued to the front to finish off the drawer. Two finger holes are used to lift and open the drawers.

95 The galley counter, which is 6 feet [1.8m] long, was made from 1/2" [12 mm], teak-faced plywood. It was coated with epoxy before being installed.

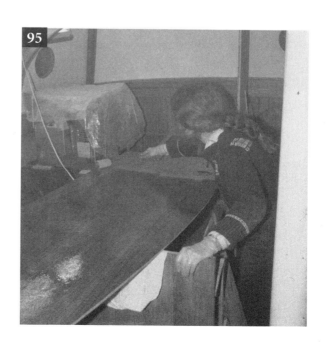

96 The cooker shelf was lined with melamine sheet. To the right in the photograph, is a knife rack and stowage for the pressure cooker lid. To the left is a laminated teak hand rail (a good place on which to hang the tea towel to dry!).

97 The Dickenson range was installed; a tiled shelf forward of it provides a good place to put hot pans. We later added removable fiddles to divide it into four, to stop pans sliding about at sea.

98 In the finished galley, there are lockers outboard of the counter, with a shelf for herbs and spices. To the left is the double sink made of teak-faced plywood, with hand pumps for fresh and salt water. A deep fiddle runs along the edge of the counter.

99 The chart table will take a standard, unfolded British Admiralty chart, and is used standing up. To the left, in the photograph, will be a stack of drawers and to the right, a large chart stowage locker.

100 The hull was lined with cork and thin veneer.

101 The framing for the doors and drawers was built in. The top of the chart locker forms a ready-use shelf for charts. Underneath the locker, is stowage space for the battery and jerricans of water.

102 A bookshelf on the heads bulkhead, over the chart table, was framed up and shelved with plywood.

103 The basic structure of the chart area was completed, ready for the drawers, doors and the melamine sheet which covers the chart table top.

104 The chart table was finished and varnished.

105 The heads compartment was fitted out next.

106 In this photograph, the heads is nearly finished. The locker fronts were simply made of butt-jointed 1 1/2" x 3/4" [38 mm x 18 mm] framework, with the joints doweled after the glue had set. A rebate was routed into the back of the framework and 1/8" [3 mm] veneers of pitchpine were glued in. The backs were planed smooth and a few air gaps were routed out. Finger holes were made in the framework.

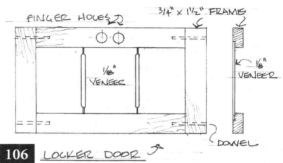

107 The heads compartment was finished and varnished. The bulkhead to the right, in the photograph, is covered in melamine sheet. The washbasin is simply a stainless steel bowl, which is emptied down the heads. The hand pump feeds from a 1 Imp gall [5 litre] jerrican which fits under the counter. The heads itself, a *Porta-Potti*, is out of sight, to the left, against the after bulkhead. Oilskins hang on hooks, next to the heads.

108 A simple framework and veneer door was fitted to close off the heads. The holes in the partition form a ladder leading to the top 'control' hatch.

109 The saloon was fitted out with settees on either side. There are lockers under the settees and lockers and bookshelves behind the back rest. The settee front, shown here, was made of plywood covered in pitchpine veneers.

110 The settee front was fitted around the floors and glued and filleted to the bottom of the hull. Two small plywood bulkheads divide the locker up and support the front and top.

111 The lockers behind the backrest are simple cave lockers with removable fiddles to hold the contents in place. The rope over the books keeps them in place in rough weather.

112 Ideally, the hull would have been fibreglassed before we turned it over, but, you may remember, a gale blew the shed down before we could do that. We fibreglassed the hull in two stages. The cabin/bulwarks and the hull topsides were done in one stage and then the bottom, of the hull, after the keel had been fitted.

The first job was to plane a vertical 'flat' at the junction of the hull and cabin sides, using a power plane. The width of the 'flat' was to be that of the rubbing stake and the 'flat' also tapers out at the bow and stern. The accurate planing of this 'flat' is very important as it defines the sheer of the hull. The hull and cabin had been given a coating of epoxy previously, and the first pass with the planer had given a very distinctive lighter line along the sheer. Being in the open, it was easy to stand back and look at its fairness. As it happened, the sheer needed lowering amidships, to produce a sweet line. This was fudged by planing more of the hull side of the 'flat' amidships, which meant that the 'flat' was not quite vertical amidships.

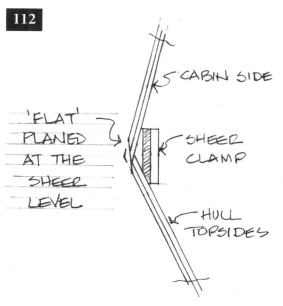

However, I feel that a sweet sheer will make or break the look of the boat, so it is very important. Once the rubbing strake 'flat' was planed on both sides, the whole hull was sanded with a coarse grit, to give a good key, and any nicks or dings in the surface were filled with epoxy putty and sanded smooth.

113 The object of sheathing the hull is to give a hard surface, and to make sure that there is a good thickness of epoxy over the plywood. We used 6 oz [140 gm] cloth and put it on in vertical strips, from the top of the cabin, down to the chine. Using a full-width roller and a paint tray, the hull was given a preliminary coating of neat epoxy resin for the width of the cloth.

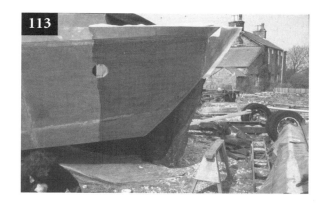

114 The pre-cut length of cloth was then hung from the deck and smoothed over the wet epoxy by hand (encased in a rubber glove!). The epoxy is sticky enough to hold the cloth in place. Another coat of epoxy was then rolled over the cloth to wet it out completely, making sure that there are no resin-starved white spots or shiny patches with too much resin. A squeegee helps here. Before the epoxy goes off, check again for any trapped air bubbles and get rid of them. The next strip was given a little overlap on the previous one. This is pleasant work if you do it on a calm day.

115 The overlaps were sanded smooth, followed by sanding the whole hull ready for the next coatings.

116 Before putting the final coatings of epoxy on, we glued the rubbing strake onto the hull. This was done in two layers, using pitchpine. The inner layer was scarfed to length and dry fitted to the 'flat' on the hull. The sheer was checked by standing back and sighting along it. The rubbing strake tapers out at each end, and this was marked on the wood. The wood was taken off, the ends shaped and the top and bottom edges planed smooth. This piece was glued in place using steel screws (which were removed when the epoxy had set). The next layer was shaped to fit, with an allowance for final finishing, and glued in place (the scarf joints being glued on the hull).

Once the glue had set, the screws were removed and dowels glued into the screw holes. The outer layer was planed flush, top and bottom, with the inner layer. The final layer of the rubbing strake was of teak, which was put on in the same manner as the second layer. This was, however, done at a later date.

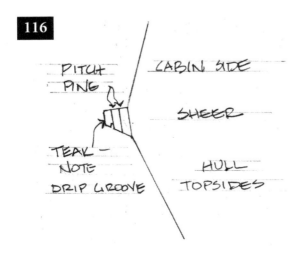

117 The weave of the cloth was filled, using pigmented epoxy, thickened with some fillers. As we planned to paint *Badger* black and cream, we used black pigment below the rubbing strake and light grey above. Filler was added to the epoxy to help fill the weave, but not so much that it would make it difficult to roll out the epoxy. The rubbing strake was coated at the same time. Two coats were used on top of the glass.

118 Jay gave *Badger* a generous boot top (which is now our waterline!) and we glued a piece of whipping twine to the hull to mark it. This has proved very successful and makes painting either the topsides or the bottom an easy job, with no masking tape required.

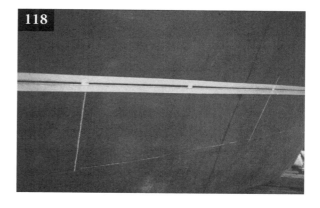

The (designed!) waterline was marked on the hull using the water gauge and measurements were taken up from the waterline at regular intervals, to get the boot top line (as shown on the plans). Masking tape was run between the marks and sighted along, to check for a fair line. A second length of masking tape was run along the hull about 1/4" [6 mm] above the first. The whipping twine (unwaxed) was taped to the bow and stretched along the hull in the 1/4" [6 mm] gap. A piece of tape kept it in place at regular intervals. Using a small brush, pigmented epoxy was brushed over the twine and the 1/4" [6 mm] gap. Once the epoxy had set, the short pieces of tape were removed from the twine, and these places were also soaked with epoxy. Finally, the masking tape either side of the twine was removed.

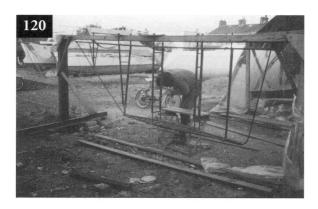

119 The ferro-cement keel has the distinct advantage of being fairly easy to build, with only a small amount of welding to do. However, the materials are not that cheap and the centre of gravity of the keel is quite high. We have since changed the keel to a cast-iron, Collins Tandem keel (a wing keel) which we bought cheap. It has a centre of gravity 12" lower than the ferro keel and reduced the draught by 7" [175 mm]. We found *Badger*, with her heavy masts, initially tender with the ferro keel. If we were doing it again, I think that we would get someone to weld up a steel keel box, which we would fill with lead to the desired weight. A flange at the top of the keel could be used to bolt it to the hull.

All that aside, here is how we built the ferro keel. Having cut out the steel for the ladder frame and keel bolts (steel threaded rod), we got someone to weld up each frame from our kit of parts. The frames were then hung from a simple overhead framework by the keel bolts and all lined up with the drawings.

120 The leading and trailing edges were wired in place. Note that the keel is suspended some distance above the ground—this made it much easier to work on without bending down all the time, and enabled us to get at the bottom of the keel.

121 The longitudinal reinforcing bars were wired on at the correct intervals. I used pliers and steel wire to do this. A loop around both pieces of metal and a few turns of the pliers, snip the excess wire off and push the turns flat. You get quite quick at it after a while.

122 The upper longitudinal should be level with the top of the ladder frames and I was careful to check again, that the curve it took matched that of the bottom of the hull. A simple plywood pattern was made to fit the curve of the bottom, where the keel would fit. Once all the longitudinals were on, the first layer of weldmesh was put on, followed by the other two—all joints staggered.

Now the tedious job starts: wiring the whole together at 3" [75 mm] centres. I found that the easiest way to get the wire through the mesh and out again was to bend it in this shape:

Holding the long end with the pliers, I pushed the loop through a hole and pulled it with the pliers. With any luck, the short end will come out of a different hole. The ends are twisted tight and any excess wire snipped off, with the twist pushed into the mesh.

123 With the metalwork of the keel completed, it was lowered to the ground onto a board, covered with plastic sheeting. A simple framework held it upright, so as to provide easy access to the top of the keel. We were lucky that a friend, Barry Whorral, had built a ferrocement boat and he gave us good advice on pouring the keel. For ballast, we were using steel punchings from a neighbouring silencer [muffler] factory. (We got a good deal because Annie worked there) Barry suggested pouring in a layer of concrete, followed by a layer of punchings and using a vibrator to mix the two together and get rid of any air bubbles. Barry was in the building trade and offered to borrow a vibrator over a weekend. We hired a cement mixer and had the cement, sand and aggregate all ready, together with a supply of water.

Barry duly turned up with the vibrator and then stayed around to check that we

were doing it properly. We soon got stuck in, with Barry doing more than his fair share of the heavy work. I mixed up the concrete and Barry poured it and the punchings into the keel. While the cement was mixing, I vibrated the concrete and punchings. Meanwhile, Annie scraped up the excess concrete that vibrated through the mesh and slapped it back into the keel. It was a full time job keeping up with the curing concrete. All was going well, with the end of the job in sight, when disaster struck. Without warning, the keel fell over with an earth-shattering thud. It just missed flattening Annie—one of the keelbolts actually grazed her arm. What had apparently happened was that the vibration had very slowly moved the keel sideways across the plank and once it was no longer perfectly vertical, it broke the supporting framework and fell over.

So, what to do now? The keel was a banana shape and would have to be rebuilt, but could we get the concrete and ballast out before it set? The cost of materials was too much for us simply to abandon it, so we set to work with a hose pipe and a length of reinforcing bar, to knock and wash out the cavity of the keel. It took hours of work, but we just managed before the concrete at the bottom had set hard. I then dismantled the keel and put the pieces to one side to think about it.

124 As it turned out, we were evicted from our building site and it would not have been easy moving the keel. When we had almost finished *Badger*, I started on the keel again and the photograph shows us preparing to pour once more. The vibrator is the red and white object behind the keel.

125 We poured the concrete and ballast as before, but made certain that the bottom of the keel could not slide and that the support framework was strong and solid. In the photograph, I am smoothing the top of the keel flush with the top reinforcing stringer—checked with the plywood pattern.

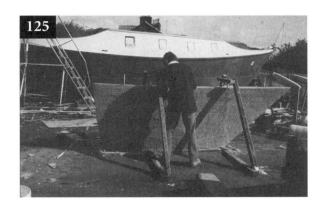

126 Once the keel was poured, it was covered with sacking and kept damp for two weeks while it cured.

127 When the keel had cured and dried out, any small holes were filled and it was faired with epoxy filler. The leading edge needed quite a bit of work to get a nicely rounded surface. A caulking tube, cut in half lengthways, made a good tool to apply the epoxy putty to the leading edge. We then covered the sides of the keel with cloth and epoxy.

128 In order to attach the hull to the keel, a plywood pattern was made of the top of the keel, with the centreline and the keelbolts all marked. Inside the hull, a centreline was drawn across the floors, using a piece of string pulled tight between the centreline marks on the bulkheads at each end of the central cabin. The plywood pattern was lined up on the centreline and the keelbolt holes positioned over the centre of the floors. The measurement between the forward bulkhead and the front of the keel was checked against the drawings. When everything was lined up satisfactorily, the centre of each keelbolt was marked on the floors. We followed the Gougeon Bros.' system to attach the keel: the bolt holes were drilled oversize, using the appropriate flat bit in an electric drill.

Building *Badger* in a yard was expensive, but it did have the advantage of having a travellift to hand for putting *Badger* on her keel. Firstly the keel had to be moved by a forklift truck.

129 The hull was lifted over the keel and lowered down onto the keelbolts for a dry run, to see that everything fitted. Fortunately, it did (the oversize keelbolt holes allowed a certain small leeway in the accuracy of their positioning— a useful advantage).

130 The hull was lifted up again and the epoxy mixed up. The insides of the holes were coated with neat epoxy, after which fillers were added, to form a non-sag mixture, using colloidal silica. This was spread over the top of the keel and slathered around the threads of the keel bolts. The hull was gently lowered down. The excess epoxy that squeezed out all round was scraped away and finished with a fillet all around the keel. Most of the bolt holes had squeezed out epoxy, but any that hadn't were filled to the top with epoxy and the oversized washers and nuts were screwed down.

131 We were very fortunate in that the travellift was not being used for any other job, so the hull was supported by the slings until the glue had set.

It is interesting to note how well this system works. *Badger* had run aground on several occasions and once went ashore in a gale, so the keel to hull joint had been well tested. When we came to remove the keel (to put on the new Tandem keel), we took all the nuts off and, using a long hole saw, made from a length of steel tubing, I cut around each keel stud down to the concrete on the keel. We jacked *Badger* up, expecting the keel to hull joint to break. It didn't: the keel came up with the boat. When there was about two or three inches of daylight under the keel, I drove an oak wedge into the trailing edge of the keel to start it. The wedge cracked the joint, but the keel needed wedging off for nearly the whole length before it parted from the hull. Pretty impressive.

132 When *Badger* was put on her keel, she was sufficiently high off the ground that we could fibreglass the bottom of the hull. We did this in the same manner as the topsides, but working overhead was not easy and it would have been much, much better to have been able to do it when the hull was upside down. If you are forced to do it this way, wear a hat.

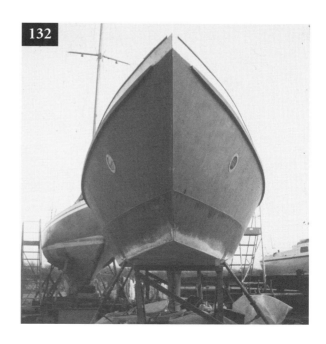

133 The rudder was built to the original design (which has since been changed) but we found on sea trials, that *Badger* was very heavy to steer and in strong winds, uncontrollable. We built a new rudder and have subsequently added a full-length skeg in front of it. I have had no experience of Jay's new rudder design, but I expect that the balance built in should correct the excessively heavy steering of the original. *Badger*'s new rudder, with the skeg, works very well, but it does have the disadvantage that the skeg can only be added to the hull after the keel has been put on. So far, we have had no structural problems with either the skeg or the rudder.

The photograph shows how we built the original rudder and the same system could be used for the re-designed rudder. As we wanted to have a trim tab for self steering, we made the rudder wider and then sawed off the trailing edge as a trim tab, after shaping the rudder blade. The correct size for a trim tab appears to be 20% of the main rudder underwater area.

The rudder was built up of sheets of plywood, glued together with epoxy, using temporary steel screws and then filling all the screw holes. The aerofoil shape was drawn on the bottom (stretched by 20% for the trim tab) and the waterline marked on. Using a power plane, the rudder blade was tapered from the waterline to the maximum thickness at the bottom of the rudder. (This is a very noisy job, because the rudder vibrates with the planing—wear ear defenders and warn the neighbours!) The veneers give a very good guide to help make the taper straight.

A line was drawn from the maximum thickness at the waterline down to the maximum thickness at the bottom. No more wood should be planed off this line. I found that the best way to mark the shape of the rudder aerofoil at the waterline was to make a pattern. Using scraps of plywood glued together to the same thickness as the rudder (ensuring that the same plywood as the rudder was being used), I transferred the aerofoil shape to the endgrain with a cardboard pattern. Where the shape crossed a veneer joint, I drew a line across the wood with a set square. Measuring from the trailing edge, I then transferred these lines to the rudder at the waterline. From the point of maximum thickness (already marked on the rudder), each line either side indicated where a new veneer should be visible. The rudder was then planed down to shape, using the veneer line marks at the waterline and the veneer lines at the bottom of the rudder. A straight edge was regularly used to check the taper. Once one side was done, the rudder was turned over and the same done on the other side.

134 After the rudder was planed to shape, a sander was used to smooth the transition from the aerofoil to the full thickness of the rudder above the waterline. The whole rudder was sanded smooth and fair with an orbital sander and the leading edge nicely rounded.

135 When the rudder was finally shaped, the trailing-edge trim tab could be sawn off.

136 The rudder and trim tab were coated and sheathed in glass and epoxy in the same manner as the hull. Cheeks were added to the rudder head for the tiller.

137 The rudder fittings are of cast bronze. I made up patterns out of plywood to fit the rudder and the stem. Making the patters was quite straightforward. An important point is to taper the fitting slightly (called the draught), to allow the pattern to be easily removed from the casting sand.

They were cast at a foundry and Annie had the holes for the pintles bored out by the machine shop at the factory where she worked. The pintle is a 3/4" [18 mm] stainless steel bar that passes through all the fittings. Great care was taken to line up the holes. The fittings were through bolted to the rudder and, on the hull, the bolts were glued into the sternpost. The trim tab was hinged onto the rudder using dinghy rudder fittings.

138 Before we had even chosen a design to build, we knew that we wanted to have junk rig. Once we had the plans for Jay's 34 ft dory, we set about designing the rig. We bought the junk rig design folios from Jock McLeod (now incorporated and expanded in *Practical Junk Rig*—published by Tiller). The design process is very straightforward and well explained. Anyone who can do simple mathematics need have no fears.

139 The ideal unstayed mast is probably one constructed of carbon fibre. To buy one ready made, would cost a small fortune and even if building it oneself, the materials are expensive. Wood seemed the cheapest option for us. Originally, we were going to buy two poles and have grown sticks, but we were offered a pile of good Douglas Fir at £4 a cubic foot and we decided to build laminated, solid masts.

140 We had hired a pickup truck to transport the wood the 100 miles back to Glasson Dock. The pickup was way overloaded and it was quite a hair-raising drive back: the brakes could barely cope with the load, going down hill.

141 Peter Latham, the owner of the marina where we were building, had gone away on holiday and it was hinted to us that now would be a good time to build the masts, in the empty boat shed. We had two weeks, so dropped everything else and got to work. The timber that we bought was 6" [150 mm] x 3" [75 mm] and about 18 ft. [5.5m] long.

The first job was to cut the scarfs to that these timbers could be joined to length. A batten was clamped on at an angle of 8:1 and the circular saw run along the batten.

142 The timbers were glued together to get the required length (the mainmast is 39 ft. [11.8m] long). The joints were dry fitted first, to check the fit and cleaned up with a hand plane as necessary. The scarfed lengths were set up level on saw horses.

143 The top of the planks were planed smooth and the next layer was glued on top, with the scarfs staggered above the lower scarfs. They were glued in place, piece by piece.

144 The problem of clamps was solved by making some simple ones, using scraps of 3/4" [18 mm] plywood and threaded rod.

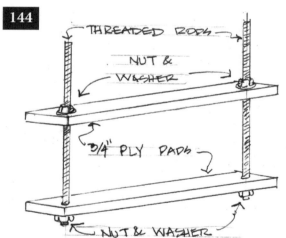

145 When everything was glued up, we had two 6" [150 mm] x 6" x [150 mm] timbers, one 39 ft. [11.8m] and one 36 ft. [11m] long. However, the masts were to be over 8" [200 mm] diameter at deck level and tapering at each end. To get the extra thickness, we glued on more pieces of wood in the required places. The photograph shows where pieces have been glued to the side; the top is being planed before gluing further pieces in place.

146 When all the pieces were glued together, the mast was sawn and planed down to a square cross section, of the correct taper—as seen on the right of the photograph, then it was marked and planed down to an octagonal section, as on the left.

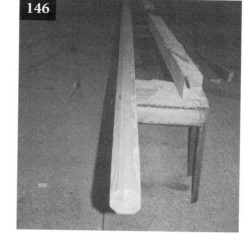

147 To mark the square section mast before planing down to 8 sides, I used a spar gauge, which is very simple to make.

The distance between the wedge points is 12" [300 mm] and the distance between the nail points is 5" [127 mm] (the nail points are 3 1/2" [89 mm] from each wedge point). The gauge is run along the mast as in the photograph. These corners are planed down to the mark.

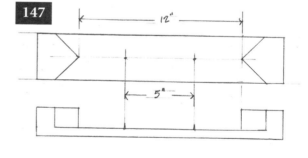

148 Once down to an octagon, the corners are taken off by plane and the whole mast rounded by eye. They were faired, using a long length of 60 grit sand paper, which was pulled back and forth around the mast. They were finally sanded along the grain with progressively finer paper. In the photograph, the finished masts are ready to be covered with glass cloth. Notice the black line on the right hand (main) mast. This is the three-strand cable for the masthead tricolour and all round white lights. A groove was cut up the mast using a router and the cable set in with epoxy filler. If you plan to put in a wire for a lightning conductor (which we did at a later date), now would be the time to do it, before the masts are sheathed in fibreglass.

149 The masts were covered in glass cloth and epoxy and painted with two-part polyurethane paint. The covering has proved most successful and the masts still look like new. They seem to require repainting about every five years, because the gloss has gone by then.

Before painting, we installed the masthead hardware. A 1 1/2" [38 mm] block of wood was glued to the masthead and a 5/8" [16 mm] galvanised eyebolt was bolted through the mast. From this is hung the halliard block. We glued two 5/16" [8 mm] U-bolts into the masthead for attaching the lazyjacks, halliard blocks and mast lift. [Illustration here]

These simple fittings take the place of the more complicated fabricated fittings shown in *Practical Junk Rig* and work well.

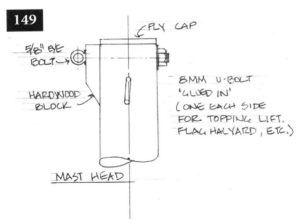

150 Our first suit of sails was made from some green, proofed canvas that I bought very cheaply. It was old stock and I suspect that the proofing had gone, as the sails rotted out after a year. The sails were simply made, using a straight-stitch, Singer, hand-cranked sewing machine. We have since used acrylic fabric (*Sunbrella*™ or similar), which is often used to make sail covers. It appears unaffected by sunlight and we have never used a sailcover. It is easy to sew and comes in a wonderful range of colours. The only drawback is that it is susceptible to chafe when it is wet, so we try and cover ropes that might chafe it with plastic tubing. We also have a thin 'battenlet' on the opposite side of the sail to the batten, to protect it from the lazyjacks. Our last suit of sails lasted 10 years and 70,000 miles.

151 *Badger* has thirteen bronze scuttles [portholes], which look wonderful and would cost a small fortune to buy. We were lucky to get them from a wrecked motorboat, that my father saw being dismantled, for £1 each. As you can see in the photograph, they were in a dreadful state.

152 After removing the broken glass, Annie spent hours grinding off paint and what was left of the chrome plating.

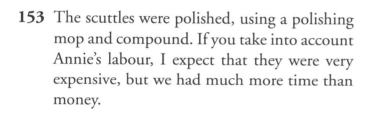

153 The scuttles were polished, using a polishing mop and compound. If you take into account Annie's labour, I expect that they were very expensive, but we had much more time than money.

154 To reglaze the scuttles, we used polycarbonate that Jack Sharples, the skipper of the Ocean Youth Club's yacht, *Francis Drake*, gave to us. This was bolted into the frame with small stainless steel bolts and silicon rubber. We had to make our own outer rings, which was simply done by making a 1/4" [6 mm] plywood pattern and having them cast in bronze at the same time as the rudder fittings.

Ten of our thirteen scuttles are opening, but we only ever open the two in the forward end of the cabin, the one in the heads and the after one in the galley. The others would be equally satisfactory if they were fixed and would be easy to make, by bolting the bronze finish ring over a piece of polycarbonate/acrylic on the outside of the hull. Then there would be only the four opening ones to buy new or second-hand.

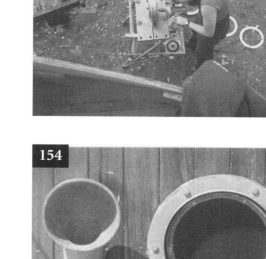

155 The cockpit had now to be finished and we decided to fit backrests, with deck lockers at their forward end. The backrests were made from 3/8" [9 mm] plywood, covered on one side with 1/8" [3 mm] strips of teak, giving a tongue and groove effect. A piece of 1 1/2" [38 mm] square teak, its face cut at the correct angle for the shape of the backrest, was glued to the deck and the backrest glued to this. A length of 1" [25 mm] square teak was also glued to the back of the cabin, to support the forward end of the backrest. Note the hole at the bottom edge of the backrest, by the cabin, to allow water to drain out through a similar hole in the bulwarks. Another piece of 3/8" [9 mm] plywood, similarly clad in 'tongue and groove' teak, was fitted between the backrest and the bulwark. The box thus formed was covered with a hinged lid. This deckbox makes a very useful seat, as well as taking the tail of the halliards and providing a store for odds and ends.

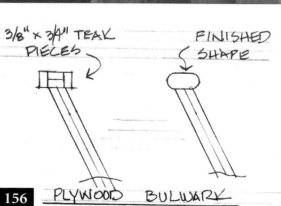

156 The capping on the bulwarks was built up from 4 pieces of teak. We found 3/4" [18 mm] x 3/8" [9 mm] to be the maximum size of teak that would bend in the two planes of the bulwark sheer. Each piece was scarfed to length and then glued on using temporary steel screws, to hold it in place. When the glue had cured, the screw holes were filled with teak plugs. After planing and sanding it fair, a router was used to round over the top edges; the lower edges were sanded round by hand.

157 We bought cast aluminium hatches at a good price. They have the advantage of being ready-made and waterproof, but unless you can get a good deal, they are expensive. They are also prone to condensation in cool climates.

Each hatch is fitted on a raised teak plinth. The plinth was planed flat, but even so, the hatches have to be put down absolutely level for the sealing gasket to work. To achieve this, the hatch was bedded onto thick epoxy to make a completely fair surface. Plastic film was used on the hatch to prevent it from sticking to the epoxy. The excess was cleaned up from around the hatch, while the epoxy was still a little soft. After it had set, the hatch was removed.

158 Finally, the hatch was bedded down on silicon rubber and held down with stainless steel screws. We used silicon rubber as a bedding compound in order to be able to remove the hatch without destroying it, and the joint has never leaked.

159 All the deck fittings, blocks, cleats, stanchion bases, etc, were screwed to the deck, the screws being set in epoxy, as in the Gougeon Bros.' book. Where there was no solid wood under the deck, we installed a piece of 1/2" [12 mm] plywood. This system works extremely well and, besides preventing any water from getting into the plywood, it is also very strong and there have been no deck leaks from the fittings. Some years ago, a 45 ft. steel yacht accidentally rammed us amidships, while we were at anchor. The whole force of the collision was taken by the stainless steel stanchion, which was bent at right angles. The cast bronze base was badly distorted and just starting to pull the fastenings out of the deck. These were 2" [50 mm], No 14 stainless steel wood screws, glued into the deck. The screws had held so firmly into the deck, that the slots on the heads had squeezed shut; they had just started to pull out. Impressive. The photograph shows the foresail sheet blocks being attached to the deck—you can see the oversize holes, filled with epoxy.

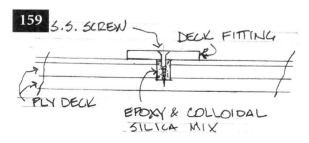

160 The table is the centrepiece of the saloon, so we tried to do something special here. Having seen an article about Vic Carpenter, in *Wooden Boat*, we decided to copy the 'grating' effect that he used for the cabin sole in one of the yachts that he built. Using solid blocks of teak and ebony, he created a solid 'grating' for the sole—the ebony corresponding to the holes.

The table is a double leaf, hinged to a centre section. The basis of it is 3/8" [9 mm] plywood and the endgrain is capped with teak strips and planed flush with the surface.

161 A border of 1 1/2" [38 mm] x 3/16" [5 mm] teak was dry fitted to three sides of the plywood, with a mitre joint at the corners. This teak border was held down by screws.

162 The teak to make the grating was offcuts from the deck. (There were plenty of 1 1/2" [38 mm] x 3/16" [5 mm] short lengths.) The majority of the pieces were cut to 3" in length as accurately as possible. The pieces that butted up to the border were 2 1/4" [56 mm] long. Everything was dry fitted to check the fit and the border pieces were glued and screwed down. Generous amounts of glue were spread on the plywood and then the teak pieces were pressed down into place, held there by gravity and friction. Instead of ebony for the 'holes,' the epoxy glue, blackened with graphite, squeezed out and filled them. Once all the pieces of wood had been fitted, the 'holes' were topped up with surplus black glue. The advantage of using epoxy instead of ebony was that it allowed for a certain leeway with the fit of the small pieces of wood.

163 The leaves and the centre section were all made and ready for finishing.

164 We made use of the boatyard's thickness planer, which made short work of levelling off the teak, to give a smooth, flat surface. An orbital sander gave the final finish, ready for varnishing.

165 The completed table's forward end is fastened to a block of wood on the bulkhead and the after end to the grab post. The leaves are attached by way of piano hinges and are supported by knees that hinge out from the bulkhead and grab post.

166 A similar system was used to make the cabin sole, with 1/2" [12 mm] plywood being covered with 3/16" [5 mm] teak in a parquet pattern. This was a good way of using up small pieces of teak.

167 *Badger*'s dinghy was a nesting one, built to the design of *Two Bits* by Danny Greene. It was constructed of plywood, using the 'stitch and glue' method.

The plans included full-size patterns for all the pieces, which were transferred to the plywood.

168 The plywood was then cut out carefully to size with a jigsaw. The hull was held together with copper wire (the stitching) to get the correct shape. The two edges to be joined, were placed face to face and small holes (just big enough for the copper wire) were drilled about every 6 inches [150 mm], about 1/2" [12 mm] from the edge. A loop of wire was then pushed through both holes and lightly twisted on the outside.

169 When all the wires were in place, they were twisted tight with pliers, until the hull is the correct shape, with all the edges fitted tightly.

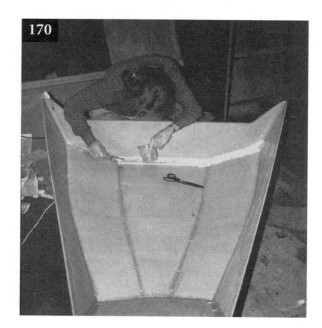

170 A fillet of thickened epoxy was used to fair in the joined pieces of plywood and a layer of glass tape used to 'glue' the joints together.

171 The photograph shows the stern section of the hull glued together by the glass tape and ready for finishing.

172 The hull was turned over and all the wires cut off close to the hull. The joints in the dinghy are sanded smooth, with the edges rounded over.

173 All the seams are covered with glass tape, following which, the outside of the dinghy is sheathed with glass cloth and epoxy.

174 When the two parts of the dinghy are completed and ready for finishing, the skeg on the stern section was glued on.

175 The forward thwart, used as the mast partner for the sailing rig, was marked.

176 Once the thwarts and mast step were installed, Annie coated the plywood with pigmented epoxy, prior to painting.

177 The finished dinghy, *Brock*.

178 *Badger* was painted and varnished, ready to go in the water.

179 On a wild and windy day at the end of January 1983, *Badger* was trundled along to the water.

180 It was a great moment as she touched the water for the first time. Annie had the champagne ready to pop.

181 The masts were stepped and held in place with wedges, over which was secured a waterproof mast coat.

182 The sails were bent on.

183 And hoisted.

184 *Badger* was now ready to go cruising. We started building at the beginning of May 1980 and set off for our first cruise exactly three years later. That was three years of hard work for the two of us. We both had jobs and spent every evening and each weekend working on *Badger*. There were no holidays and very few days off. As I worked at a teachers' training college, I had quite long 'holidays,' which helped progress a lot. By the end of three years, we were about 'built-out' and I don't think that we could have kept up the single-minded pace much longer. We often thought that had we not finished *Badger* in three years, it would have taken five! Another interesting point is that after launching and sailing *Badger*, it took quite a long time to get out of boatbuilding mode—it took some time to settle down to cruising. I'm sure that if someone had offered a good price for *Badger*, we might have been tempted to sell her. All in all, though, it was a completely worthwhile project and it has been very satisfying to voyage in a boat built with our own hands.

Postscript

In point of fact, after cruising her over 100,000 miles, Pete and Annie succumbed to the lure of an attractive offer for her and **Badger** passed on to new stewards.

The Most Economical Offshore Cruisers

By Jay Benford

The original versions of these sailing dory designs were done in the 1960s. I laid them out in trying to find the simplest sort of capable boat that I might be able to afford. Publishing them before I had a chance to build one—or could afford one—we've ended up with designs for several different sizes and with hundreds of them under construction and sailing all over then world. They are being built both in home and professional shops. And I still don't own one. . . .

Each of these sailing dories is designed with a simplified construction plan, using sheets of plywood from the bottom up. The interior bulkheads are the only transverse frames in the vessel, with the addition of the floors in way of the keel. The keel (lead or concrete and scrap metal, depending on the version) is usually a sub-assembly bolted on prior to launching. The longitudinal framing consists of the chine, clamp, and raised deckhouse and/or house framing; there are no ribs to contend with. There are no deck beams, except for a temporary framework not finely fitted for the multiple layers of plywood above. The straight lines (see the lines plans in the following chapters) means extremely easy lofting; a professional shop can loft one of the larger boats in a few hours! The straight lines also mean that there are a lot fewer curves to contend with in outfitting; thus it is much easier to attach interior joiner work than in a typical (more curvaceous) hull.

The first of the 36-footers was built in under 13 months by a shoemaker in Canada. He had never built a boat before, and didn't even have any boatbuilders with whom he could consult within his vicinity, 1200 miles from the coast. One of the first 19-foot Gunkholers (See our **Catboats** book) was built by professional boatbuilder Paul Miller, of Port Hardy, British Columbia, in a little over eight weeks. Thus these vessels make an ideal project for the amateur builder or for the professional yard who can set up for plywood production on a practical basis.

Though these vessels are flat-bottomed, in dory fashion, when heeled over they present a "V" to the water, and thus move along at a good clip. We've had the pleasure of experiencing just how well this works when Fred Schreiner took us out for a sail in his 36' Sailing Dory, **Donna**, on San Juan Channel. Fred later sailed **Donna** from Vancouver, Canada, to Mexico, as well as to Hawaii and back to Victoria, B.C. He remarked that she tacked very easily and would hold her course well. To quote him:

"It really surfs down those big, Northwest swells with a good wind. Close hauled it steers itself. At other point I use a steering vane. It is actually a trimtab on the large rudder with a vane and adjustments on top. It really works good. Often I do not touch the tiller for days on end, that is, if the wind stays that way."

They are thus lively and fast, with lots of sail power to carry their light, but well-ballasted hulls in a wide variety of wind conditions. The chine on each is well immersed at the bow, to minimize or do away with pounding at anchor. Each fin keel and rudder are NACA foil sections. The generously large rudders maximize ease of control and minimize chances of stalling the rudder and broaching (which is common to the small racing spade rudder types). Each of the sailing dories has a diesel engine for auxiliary power. The small (10 to 20 hp) amount of power called out is quite adequate to push these vessels to hull speed, and they have the additional advantage of being simple, safe and low in fuel consumption. Oil or wood stoves are called out for all the dories, as they make for dry heating and pleasant cooking, although other fuels could—and should—be used for cooking in the tropics.

ALTERNATE CONSTRUCTION METHODS

We're asked occasionally whether the dories could be built in other materials. The answer is a qualified yes. We've done alternative construction plans for the 36' and 37½' dories for building in aluminum. It would be possible to do the smaller dories in aluminum, with similarly revised construction plans.

Ferro-cement is too heavy a method for these boats, and not suited to the flat panel type of construction in lighter weights.

Steel would be possible, but it would substantially increase the displacement because it is a heavier material. This would negate some of the advantages of these designs, so we don't recommend it, even though it would be possible. However, if a steel version were to be designed, it would require a hull form that is beamier and more burdensome to carry the signficantly heavier structural weights. This would in turn lead to the need for more sail area and a larger engine. In designing the steel 36' Power Dory version of the 36' Sailing Dory, the additional structural weight equaled the ballast on the sailing version. . . .

Airex® cored fiberglass or C-Flex fiberglass could be done, but this is not the best type of design for one-off fiberglass building. The 'glass method that would work best is to make a large, smooth surfaced table. This would then be waxed, sprayed with gel coat and then the 'glass laid up on it. The large flat panels thus created would be put in place over the mold frames for the dory, and trimmed to fit. The corners would then be 'glassed together, suitably reinforced and then gel coat repaired in way of the seams.

Plywood and epoxy is still the best choice for construction of the dories. It uses simple, inexpensive materials, and the final structure provides its own insulation. It is also easy to add items of outfit to the boat, by simply gluing or screwing on pieces inside, or bolting on deck hardware, It is my own first choice for building these boats.

A wooden boat is at least as strong as the other materials. In addition she's warmer, prettier, and often better finished. Because she is all of these things, her owners feel greater pride in her and they usually pay her the attention she deserves in the way of handling and maintenance.

And what of maintenance? The prospect of "rot" makes us all concerned. In reality, the major causes of rot are poor construction techniques and truly gross neglect. To quote Paul Miller, who's had years of experience repairing, renovating, and rebuilding older wooden vessels, "Properly built boats rarely rot." They should be designed and built with good ventilation (so the wood can breathe) and the end grain should never be exposed to weather. Aside from those few key points, maintenance of a wood boat is limited to painting. Paul points out that "today's linear polyurethane paints, applied in conjunction with the proper quality epoxy sealers can produce a finish that will outshine and outlast the gelcoat finish on fiberglass hulls." Epoxy sealers are also a boon to the endurance of bottom paints.

Marine grade plywood is specified in our plans. Hull, deck, and cabin are assembled with bronze screws and epoxy glue, then completely sheathed with fiberglass cloth set also in epoxy resin. Note that the sheathing cloth is not set in polyester resin, for polyester does not adhere as well as epoxy, nor does it last as long. We have seen too many plywood 'glass sheathed boats with their outer layer peeling off. The plywood-epoxy cloth method of wooden boatbuilding (much like cold molding, only with thicker or preglued "veneers") results in a strong, totally watertight, monocoque shell.

When the time comes to outfit the interior, plywood is definitely easier to fasten to than many other materials. Additionally, the inside of the hull needs no further insulation, and as it's pretty to the eye, it needn't be covered up for any aesthetic reason. As mentioned earlier, and as Paul Miller enjoys pointing out, "the hard chine dory hull form makes for many straight line fits, with very few difficult curves." All of these points add up to timesaving without sacrificing beauty, which in turn translates to a lovely boat at low cost.

Rigs

There are several different rigs on the dories. Each has its virtues. You should select the one that best matches how you will use yours.

The lug ("junk") rig permits handling all sail evolutions from the cockpit. It is quickly and easily reefed. It does have a bit more weight aloft and it is not as easy to add light air sails on it as on the cutter.

The ketches were designed for ease of balancing the helm and keeping the course with a minimum of effort. They are mostly gaff rigged on the mains to keep the spars down. This keeps the center of effort low, so the heeling moments are lessened, permitting sailing at lower angles of heel.

The cutter and sloop rigs have the tallest masts. The longer luffs will give better windward ability. The height will permit flying larger light air sails, optimizing performance. These rigs are the most conventional, with a wide selection of hardware and parts available for them. They will probably enjoy the best resale value, being the most widely used and understood rig.

The Dory Hull Form and Stability

Some alarming tales are heard from time to time about the dory type of hull being "tender" — that is, heeling easily initially and then stiffening up as the deck edge approaches the water. This has been the case with the rowing dories for so long that it has become an article of faith among many that **any** dory will be tender. There are still a number of rowing dories being built where this is still the case, as this also makes for an easily rowed boat.

However, our modernized versions have relatively wider bottoms and are a bit fuller aft to give more initial stability and make for better sailing performance. This updated hull form, combined with a 40 percent ballast ratio carried low on the keel, makes for a pleasantly stiff boat. The additional design data, simplified lines plan, and the stability curve of the 36-footer on these boats should serve to silence the skeptics. Further testimony on this comes from Fred Schreiner, while sailing his 36-foot sailing dory ***Donna***, enroute to Mexico:

"Sailing in the ocean with the powerful winds and large swell is different than in the protected waters here. The boat is easy to steer, turn and keep under control. It reacts quickly and I can steer in any tight spot in port. The best of all is when I have back out in reverse gear. It goes where I want it to go. Sometimes I sailed up to the mooring buoys or piers with very little wind and hit the buoy or pier dead on, because the boat steers so precisely. The sails come down like lightning. But it takes a good crew. The boat surfs with a fair wind down those big rollers. Another good important item is the double ends. When I was off shore in the storm, I think it must have been Beaufort Scale 8, not one breaker entered the boat's cockpit. The sea was following. When it got too tough and rough we let the boat drift. The motion was violent, and the noise terrible, but the boat took on waves just like nothing. I write this because I sailed on production boats in moderate seas and they jump around. I wonder what they do in force 8 winds. When I sailed and motored around Cape Anguello in a rough wind and confused sea another ketch got dismasted and drifted around for three days till the Coast Guard pulled them into port. At times the waves were 16' high. Then the boat fell off then wave top down in the trough with an awful shudder and bang. The bowsprit poked into the next wave already. But the boat withstood it all. No damage to it at all. I am well pleased with your design."

We're quite proud of the performance and sail carrying power we've designed into these boats, as we know this gives pleasure and a feeling of security to those aboard. Try it. . . .

22' Sailing Dory
Design Number 17
1965

This is the first of the five dory designs done originally for the late Captain Jim Orrell, who ran Texas Dory Boat Plans. Having left my apprenticeship with John Atkin a year before, where we'd done some designs for Texas Dories, I sent an idea to Texas Dories and got a go-ahead to create the plans. These drawings, and the preliminary below, are the result.

Some evolution from the preliminary conceptual drawing is evident; a revised rudder, rig and provision for more than one crew to sleep aboard. Most of us don't cruise alone. . . .

The structure is different than our later dory designs in that it has actual framing beyond the bulkheads and furniture. I'd be tempted to modify the structure to be able to leave out the framing and have the smooth interior to make outfitting easier. Also, the headroom is low, particularly if the sole is installed. It would be interesting to look at creating a raised, flush deck version and see if this was a way to have more useful room inside and not spoil her looks.

Looking at her rig now, I'd look at adding a short bowsprit for pushing out a small roller furling jib. This would also give a good place for an anchor roller. The other thing I'd think about would be adding a little more depth so that the accommodations would have better headroom in the cabin, perhaps by adding height to the cabin. And I'd like to see provision for an outboard in a well or a saildrive engine.

All in all, she should be a lively little cruiser and a fun way to go exploring.

Table of Particulars:

22' Sailing Dory	Imperial	Metric
Length Overall	22'-0"	6.71 m
Length Datum Waterline	18'-0"	5.49 m
Beam	7'-0"	2.13 m
Draft—fin keel	2'-6"	0.76 m
Freeboard, Forward	3'-0"	0.91 m
Freeboard, Least	1'-8¾"	0.53 m
Freeboard, Aft	2'-2"	0.66 m
Displacement, Cruising Trim*	2,375 lbs.	1,077 kg
Displacement-Length Ratio	182	
Ballast	600 lbs.	272 kg
Ballast Ratio	25%	
Sail Area, Square Feet	183	17.0 m²
Sail Area-Displacement Ratio	16.45	
Prismatic Coefficient	0.57	
Auxiliary Horsepower	5	
Headroom	3'-4"	1.02 m

*__Caution__: The displacement quoted here is for the boat in coastal cruising trim. That is, with the fuel and water tanks filled, the crew on board, as well as the crews' gear and stores in the lockers. This should not be confused with the "shipping weight" often quoted as "displacement" by some manufacturers. This should be taken into account when comparing figures and ratios between this and other designs.

Building *Badger*

Notes to Builders (1965)

Setting Up: After you've completed the lofting, and have some materials in hand, the next step is getting prepared for the construction. A good construction site must be decided upon, preferable one that would provide protection from the weather, and provide a good base for a level reference frame. Most two car garages will be large enough to built the boat in, with the centerline of the boat set on a diagonal inside the garage.

From the lofted lines, the size and shape of the frames and internal structure can be determined. Be sure to subtract the thickness of the planking from the lofted lines, as the offsets are to the outside of the planking. The required bevels can be readily picked up from the lofting.

The boat can best be built, up to the point of putting in the interior and decking, by setting up upside-down. If you extend the topside frames all a set distance beyond the waterline and somewhat beyond the sheer, these can all be affixed to a level floor or reference base. Then, when the boat is turned over for decking, the frames are cut off at the required length, and the deck beams put in.

The deck has a crown of 3" in 7'-0", and the deck beams should be sawn to this camber. The ends of the deck beams should be notched to fit around the harpin, and screw fastened to it. The same process should be carried out where they meet the carlin. The only deck beam not sided ¾" is the one at the forward end of the cabin. This one is sided 2", and is intended to be a fastening base for the ¾" mahogany forward end of the cabin.

The stem is made up in two pieces, the outer one being ¾" x ¾" and fastened on after the forward ends of the planking have been trimmed off flush with the inner stem. The inner stem has a depth, normal (900) to the edge, of 3" at the top and increasing to 4" at the bottom.

For those who are concerned with the ultimate longevity and care to put in a little extra effort, there are a few details that will help. One, it to bevel the top of the inner chine so that it will not trap moisture on it when at rest. Also, be sure to use bronze or Everdur fastenings when fastening the parts together—brass will not hold up in salt water. And be sure to use good quality marine bedding compounds when putting pieces together, and good marine paints in several coats over the whole of the boat.

Outboard Well: The outboard well is shown as an option, and it not a real necessity in areas where there is good and constant wind during the sailing season. If it is installed, be sure to securely fasten it to the bulkhead at station 11, as shown, and also to the frame at station 10. The cutout at the bottom should be just large enough for the cavitation plate of the engine to fit through. The piece cut out should be saved and fastened to a slightly larger piece to provide a flush insert for when the engine is not in place while sailing. If your engine tends to pump water into the well while powering, a drain hose of about 1½" diameter should be fitted about 2/3rds the way up one side and draining down and overboard through a thru-hull fitting.

A good alternative for auxiliary power would be to carry a sculling oar. If you fitted an oarlock on the starboard corner (port if you're a left-handed sculler) of the transom, this would be about all the installation work you'd have to do for your "auxiliary".

Open Version: I've drawn an open version for those who want to carry a couple more people along just for daysailing. I would give it a lot of serious thought before deciding to build this version. The cockpit of the cabin version will hold four in good comfort, and the cabin provides something of a shelter in poor weather and a place to change into and out of swimming gear.

Misc.: There are some very good advantages to full length planking which should be considered very seriously. The full length pieces eliminate the time and effort spent in butting the pieces together and putting in butt blocks and getting the joint watertight. Also, one does not have to spend a lot of time figuring out where to put the joints so they will be well staggered, and the lack of these joints makes for a stronger hull. The sides can be gotten easily out of 24' lengths, and the bottom from a 20' piece. You may have to do some talking and looking, but these things can be obtained. *(Scarfing the panels together yourself is readily done now.)*

Sails are power—have some good ones made!

This photo shows one of these 22-footers built in Germany, tucked away in her storage area.

Below is the preliminary design concept and the following pages have the working drawings, as done in 1965.

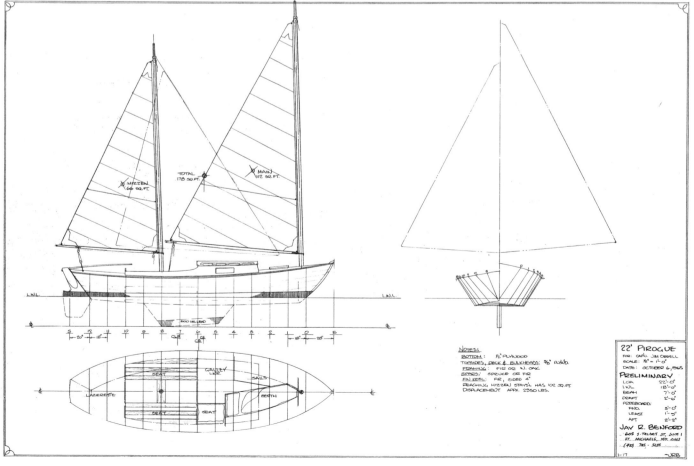

Building *Badger*

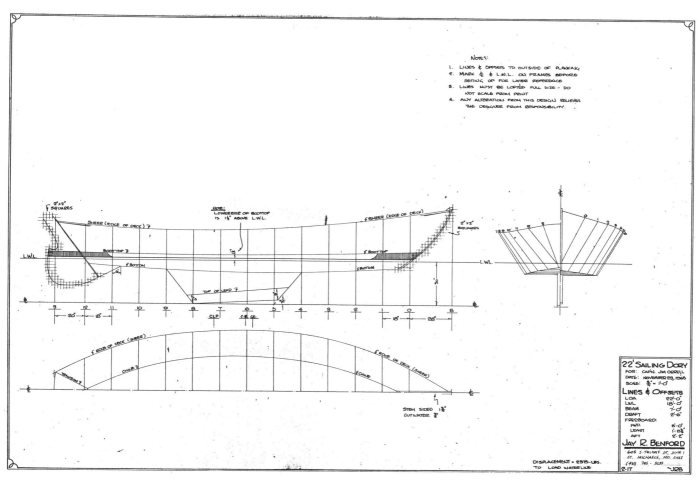

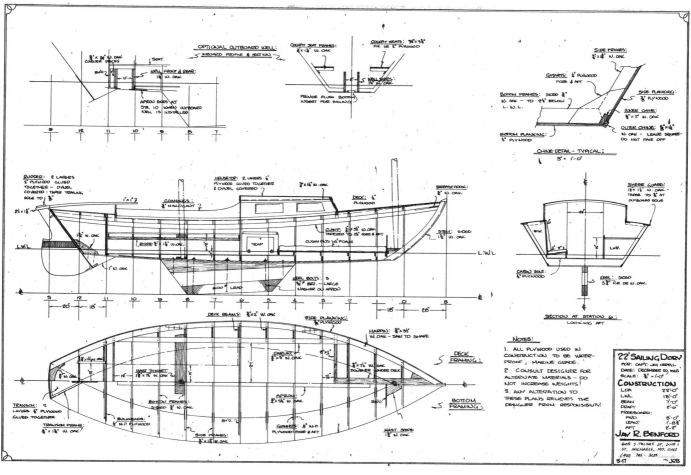

Building Badger

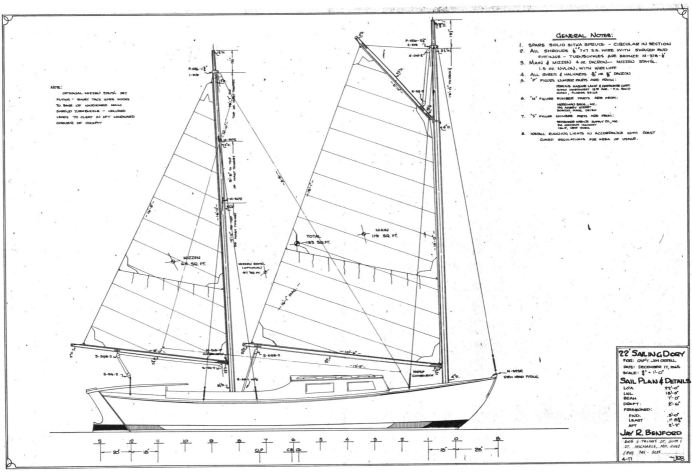

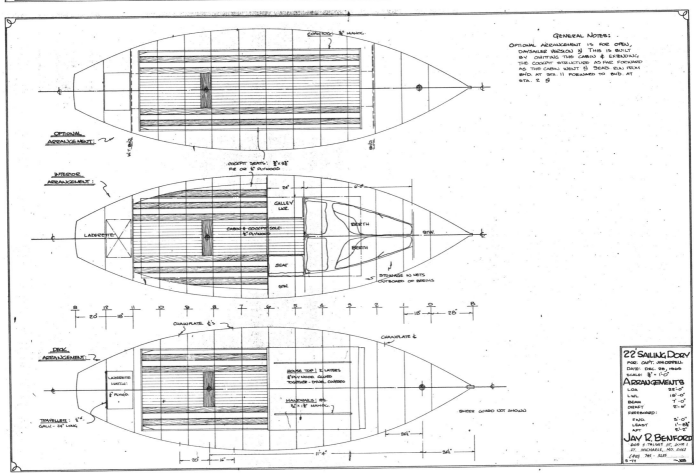

26' Sailing Dory
Design Number 21
1965

This is the second dory design done for the late Captain Jim Orrell of Texas Dory Boat Plans. This one is more like a sharpie than the 22, developed as I was looking into ways to gain stability and volume in the cruising accommodations. This was done by making the bottom relatively wider and having less flare in the topsides.

The Revision A drawing has a modification to the skeg for inclusion of an inboard engine. And a modified interior layout with a quarter berth for an additional crew member to come along.

The structure also has actual framing beyond the bulkheads and furniture; something we've gotten away from in our later dory designs. Revising the scantlings to eliminate these frames would be something worth considering, as the resulting gain in interior livability would be appreciated. And, the outfitting would be easier with a smooth interior.

Like a lot of older designs, the engines called out are probably no longer current production items. They could be readily replaced by a small (5 to 10 hp) diesel. The same situation probably exists with some hardware specified no longer being available with companies no longer producing them or out of business.

Table of Particulars:

26' Sailing Dory	Imperial	Metric
Length Overall	26'-0"	7.92 m
Length Datum Waterline	24'-0"	7.32 m
Beam	7'-0"	2.13 m
Draft—fin keel	3'-0"	0.91 m
Freeboard, Forward	3'-9"	1.14 m
Freeboard, Least	1'-9"	0.53 m
Freeboard, Aft	2'-6"	0.76
Displacement, Cruising Trim*	4,250 lbs.	1,928 kg
Displacement-Length Ratio	137	
Ballast	930 lbs.	422 kg
Ballast Ratio	22%	
Sail Area, Square Feet	281	26.11 m²
Sail Area-Displacement Ratio	17.14	
Prismatic Coefficient	0.70	
Auxiliary Horsepower	5 to 10	
Headroom	4'-4"	1.32 m

***Caution**: The displacement quoted here is for the boat in coastal cruising trim. That is, with the fuel and water tanks filled, the crew on board, as well as the crews' gear and stores in the lockers. This should not be confused with the "shipping weight" often quoted as "displacement" by some manufacturers. This should be taken into account when comparing figures and ratios between this and other designs.

1966 Additional Architect's Notes (Updated 2010):

Building Time—professional estimates in 1966 were for about 700 man-hours—prior to painting and rigging. Scale this up considerably if you've never done it before! . . .

Sheet 2-21 Lines & Offsets:

Side curvature for transom should be developed on the job. Continue as much curvature as needed to keep plywood from buckling.

Sheet 3-21 Construction & Arrangement:

Anyone desiring to use something other than lead for ballast—iron, concrete and iron scrap—should consult with the designer for size and amount. About 300 to 500 pounds of inside ballast will be needed in most cases. This should be distributed between stations 4 and 8 under the floorboards.

Setting Up—This boat is best built upside-down. If you have a level floor to build on, extend the frames out to where they are 4' above the waterline, and fasten their heads to the floor, trimming them to the final length after rolling over. If you don't have a level floor, try setting a couple 2x8's 4' apart with their tops parallel (and level) for a base to work from.

Optional Engine—the installation drawing for an inboard engine on sheet 3-21 is for those who feel they must have an engine. This boat will sail very well without one, and I would not put one in unless you have a special need for it, such as a long tricky harbor entrance. Other suitable engines would be in the under 7 hp range, preferably a diesel.

Bottom can be two layers of 3/8" plywood with seams staggered and the seam batten can be omitted. Topsides, decks and housetops to be two layers of ¼" plywood. Glue and clout or Anchorfast nail the layers together. Framing and structure may be Alaska (yellow) cedar or Douglas fir as suitable.

Plywood to be MDO surfaced if painted or mahogany faced if varnished.

A water tank should be fitted (about 5 to 6 gallons) under the counter that the sink mounts in for supplying the sink.

ERRATUM: Clamp isn't shown on the section at sta. 10—please don't leave it out in building her!

The cockpit can be improved by raising the sole to 9" above the LWL and making it self-draining. Raise the seats to 18" above LWL, enclosing the space under them for storage and put sloped gutters under the seat hatch openings to drain the water back into the foot well.

Rudder can be three layers of 3/8" plywood glued together, tapered to 5/8" thickness at trailing edge. Minimum two sets of pintles and gudgeons.

Area just forward of propeller aperture to be faired down for better water flow to prop.

Sheet 4-21 Sail Plan & Deck Arrangement:

ERRATUM: Shroud and stay diameter should read 5/32" diameter rather than 5/16".

Spars to be solid Sitka spruce

Least expensive end swages for the standing rigging are Nicopress sleeve swages—possibly available from your local telephone company. Wilcox-Crittenden "Norseman" turnbuckles will save swaging the lower ends of shrouds and stays.

Provide four 8" cleats, two forward & two aft, plus chocks if desired.

SAILS ARE POWER—have some good ones made! A slight extra investment here will pay many dividends in performance in longevity.

Sheet 5-21 Revision A:

Additional suggested gear to have aboard:

- 18 pound Danforth anchor.
- 300' of ½" Nylon for 200' of rode (with 20' of ¼" chain leader), and four 25' mooring lines.
- Leadline & Boathook
- US flag and guest flags for countries to be visited and flag halyards.
- Cockpit awning; to extend well forward of companionway opening.

Shown below is the Preliminary Study for this 26-footer. You can see how this concept is carried through in the final version of the design on the following pages. Also, the body plan shows what the sails would look like if running straight down wind with them all out 90-degrees from the centerline. The shadowing of the two on one side would be the result of doing this in real life, with the forward most one not working as hard as the others. Better to be on a broad reach and have all them pulling well.

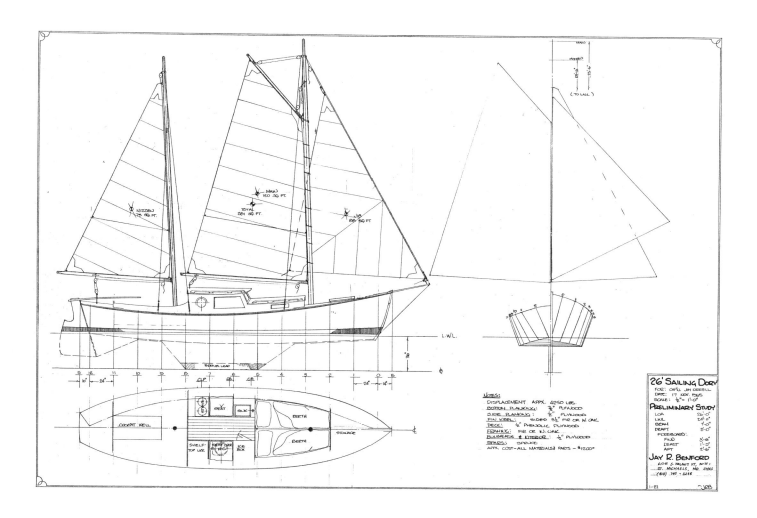

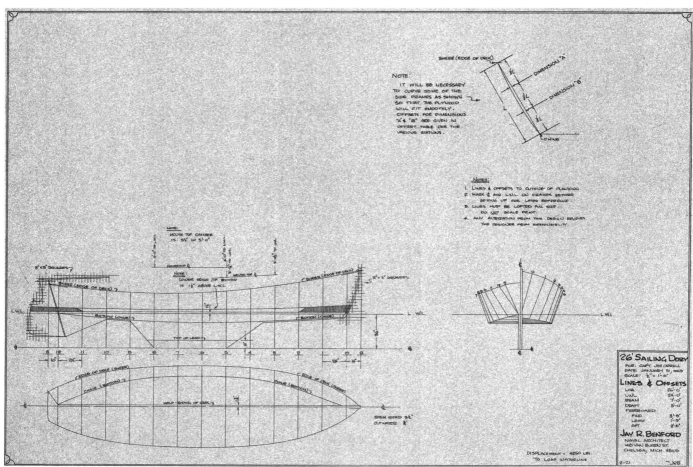

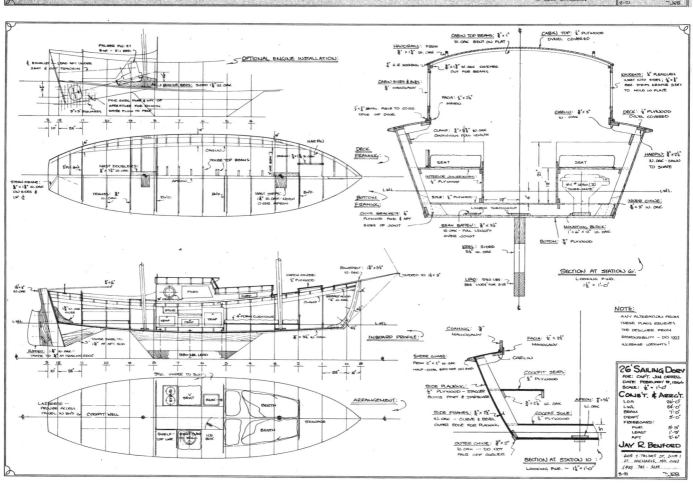

Building *Badger*

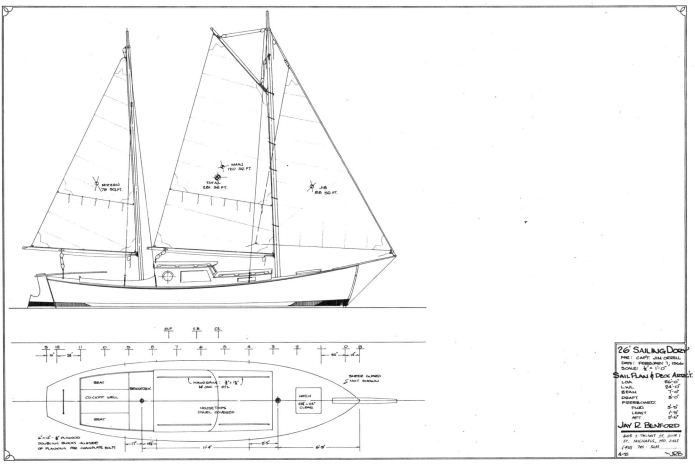

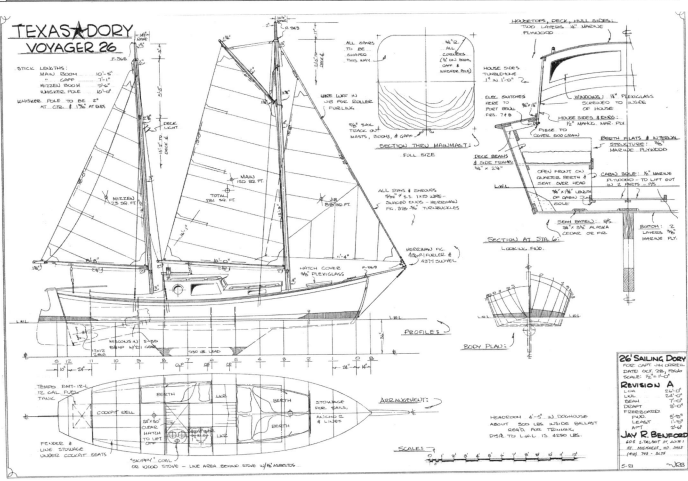

27' Texas Dory
Design Number 24
1965

This theory behind this preliminary design was to develop a double chine 27' Sailing Dory. She has the same narrow 7' beam that the 26' Sailing Dory does, but reaches her maximum beam closer to the waterline, meaning that the gain in stability as she heels will be happening at lower angles of heel.

Overall, I guess I'm just as glad that we did not follow through with this to completion. File this one under food for thought. The improvements I'd like to make to her would include more beam, more depth and updated accommodations.

I much preferred the approach on design number 33 for a beamier and roomier yet simple 27-footer; which is also still an unfinished design. You'll find her a few chapters further into this book.

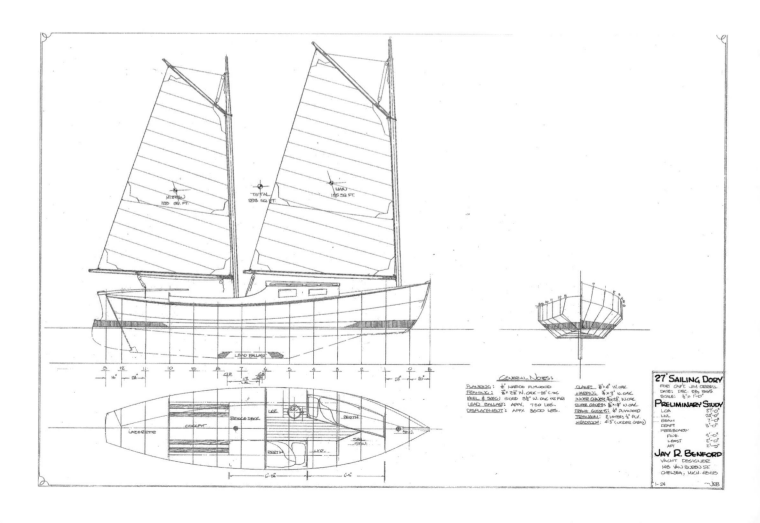

26' Saint Pierre Dory
Design Number 27
1965

This is the third dory design done for the late Captain Jim Orrell of Texas Dory Boat Plans. He had sent me some drawings of a traditional St. Pierre dory and asked me to create a set of working plans for it. He wanted to stay close to the traditional appearance and provide plans for people to build them. He also asked for 23' and 19' versions and these are shown in the following chapters.

The St. Pierre Dory type is most readily identified by its distinctive great and sweeping sheer. The late John Gardner, who wrote for Maine Coast Fisherman, then National Fisherman—and many other magazines—wrote praising this type of dory highly. His books, many from his years on staff at Mystic Seaport, includes **The Dory Book**, which has information on this type and many others. These books have a wealth of useful information in them for those who want more historical context and who are interested in how to build the varied historical small craft types.

Built and originated in the St. Pierre et Miquelon Islands, off the coast of Newfoundland, the St. Pierre dories were originally built as working vessels, putting to sea in all sorts of weather, and bringing home their catch to market. They are powered easily with modest sized engines, and some were even rowed and sailed. Many had a device for hauling the propeller up into a watertight box in the bottom, making beaching the boat safe for the prop.

The Preliminary Study gives a good view of the direction the design would follow. The subsequent drawings have the Lines and Offsets,

Table of Particulars:

26' St. Pierre Dory	Imperial	Metric
Length Overall	26'-7"	8.10 m
Length Datum Waterline	20'-6"	6.25 m
Beam	7'-7"	2.31 m
Draft	2'-0"	0.61 m
Freeboard, Forward	4'-4"	1.32 m
Freeboard, Least	1'-9"	0.53 m
Freeboard, Aft	3'-5½"	1.05 m
Displacement, Cruising Trim*	4,000 lbs.	1,814 kg
Displacement-Length Ratio	207	
Sail Area, Square Feet	39	3.62 m²
Prismatic Coefficient	0.59	
Auxiliary Horsepower	7 to 10	
Headroom	4'-6"	1.37 m

*Caution: The displacement quoted here is for the boat in coastal cruising trim. That is, with the fuel and water tanks filled, the crew on board, as well as the crews' gear and stores in the lockers. This should not be confused with the "shipping weight" often quoted as "displacement" by some manufacturers. This should be taken into account when comparing figures and ratios between this and other designs.

giving the geometry of the boat, the Construction with details on how she is to be built, and the Profile & Arrangements with her appearance, interior layout and rigging notes.

In addition to the original planked version, there are details for a plywood variant in these plans too. Looking at them today, they seem to be of the heavy, workboat scantling approach. I suspect that we could, in designing a yacht version, create a lighter structure that would allow for carrying more in the way of interior accommodations and gear and still provide a strong and rugged structure.

The little riding sail is intended to help keep her head to the wind when hauling lobster pots in a swell, and to help dampen the motion in a seaway.

The outboard well, shown on the Lines plan is an alternative for those wanting to use an outboard engine. The typical hazard with an outboard in a well is the accumulation of exhaust fumes boiling up inside the boat. It can also cause a problem is the boat is heavily laden and water floods from the well into the boat! Because of this danger, it would be good to fit a secondary compartment around the well that is much taller and that would limit any flooding to the immediate area at the well. Best to be cautious, stay afloat, and complete every voyage you start. . . .

Later on in this book, there is a chapter on our design for an even larger version, the 32' St. Pierre Dory, which was created to be a commercial West Coast fisherman.

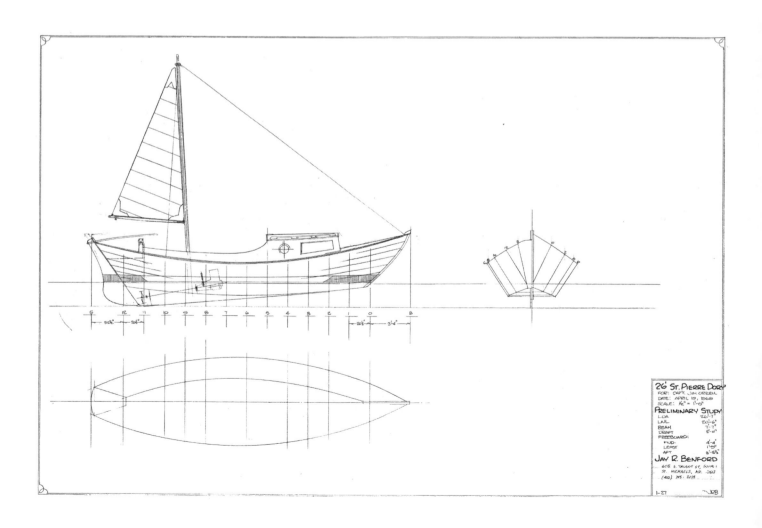

Building *Badger*

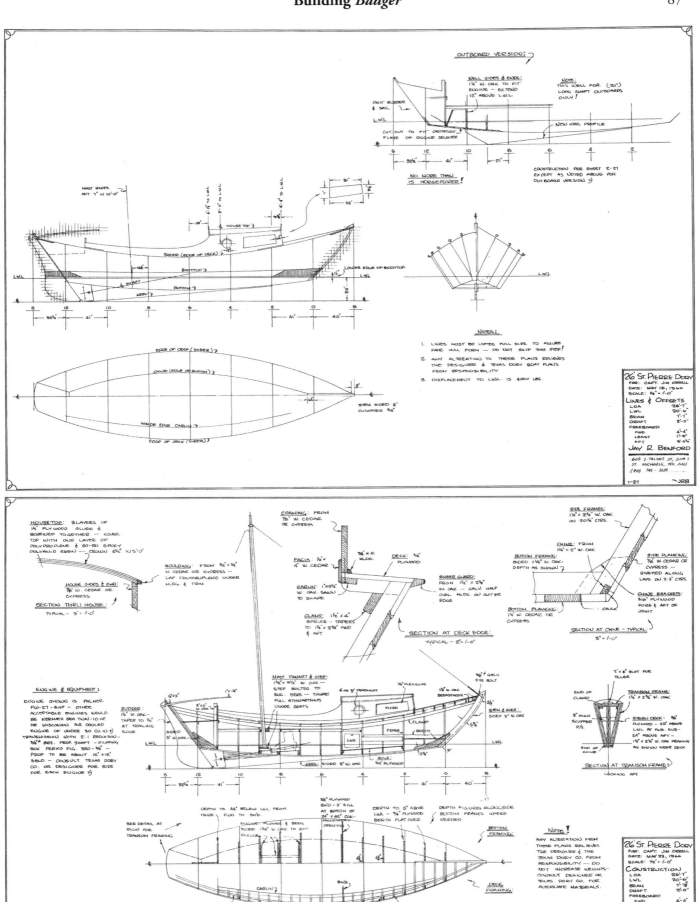

Building Badger

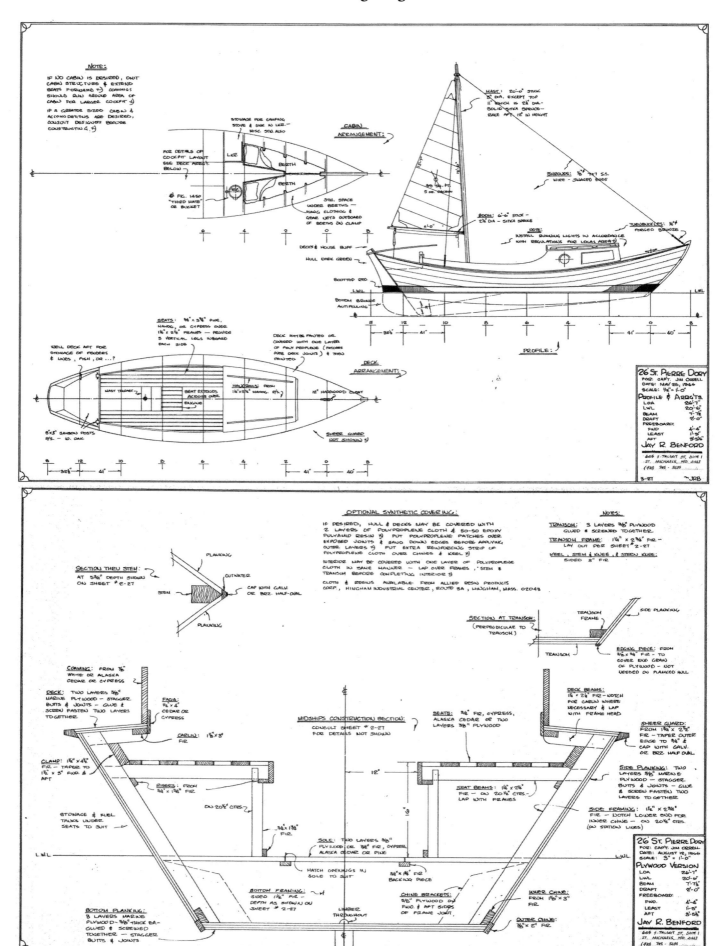

23' Saint Pierre Dory
Design Number 29
1965

This is the fourth dory design done for the late Captain Jim Orrell of Texas Dory Boat Plans. After doing the 26' St. Pierre Dory, we scaled it down three feet to create this one. The 23 is only shown as an

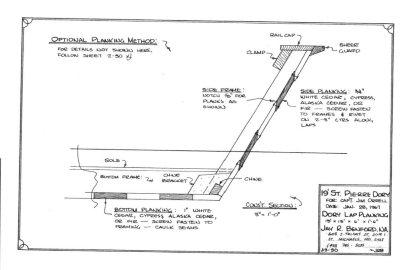

open dory, with seating for a big picnic party to enjoy day trips. It would be possible to create a little cuddy or add a melon hood to have some shelter aboard. This would provide a space to have some privacy for using a head, or changing in to or out of swimming gear.

A more contemporary 4 to 8 hp diesel would be the power of choice—unless you're into rebuilding antique marine engines, like the ones noted on the drawings....

The unfinished Dory Lap Planking drawing is like the section on the preliminary drawing, and is another approach to planking the boat. See the comments on the 19' St. Pierre about the final design all being on one sheet.

Table of Particulars:

23' St. Pierre Dory	Imperial	Metric
Length Overall	23'-0"	7.01 m
Length Datum Waterline	18'-0"	5.49 m
Beam	7'-0"	2.13 m
Draft	1'-8"	0.51 m
Freeboard, Forward	4'-0"	1.22 m
Freeboard, Least	1'-9'	0.53 m
Freeboard, Aft	3'-2"	0.97 m
Displacement, Cruising Trim*	2,800 lbs.	1,270 kg
Displacement-Length Ratio	214	
Sail Area, Square Feet	122	11.33 m²
Sail Area-Displacement Ratio	9.83	
Prismatic Coefficient	.57	
Auxiliary Horsepower	4-8	1.37 m

*****Caution**: The displacement quoted here is for the boat in coastal cruising trim. That is, with the fuel and water tanks filled, the crew on board, as well as the crews' gear and stores in the lockers. This should not be confused with the "shipping weight" often quoted as "displacement" by some manufacturers. This should be taken into account when comparing figures and ratios between this and other designs.

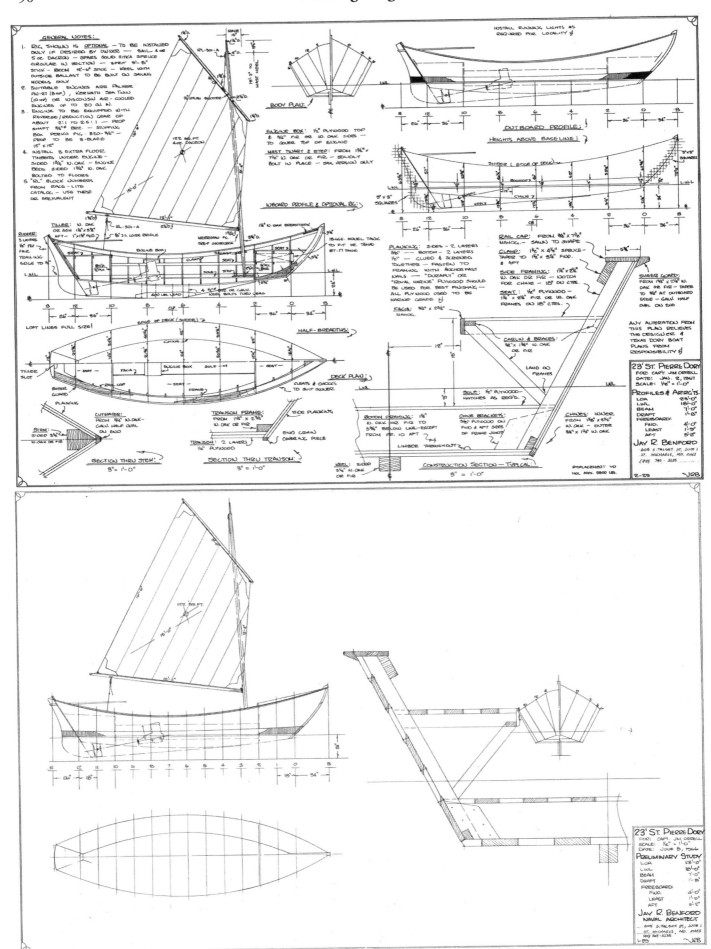

19' Saint Pierre Dory
Design Number 30
1966-9

This is the fifth dory design done for the late Captain Jim Orrell of Texas Dory Boat Plans. This one, while the smallest,

Table of Particulars:

19' St. Pierre Dory	Imperial	Metric
Length Overall	19'-0"	5.79 m
Length Datum Waterline	15'-0"	4.57 m
Beam	6'-0"	1.83 m
Draft	1'-6"	0.46 m
Freeboard, Forward	3'-3"	0.99 m
Freeboard, Least	1'-6½"	0.47 m
Freeboard, Aft	2'-9"	0.84 m
Displacement, Cruising Trim*	1,750 lbs	794 kg
Displacement-Length Ratio	231	
Ballast	275 lbs	125 kg
Ballast Ratio	16%	
Sail Area, Square Feet	98	9.10 m²
Sail Area-Displacement Ratio	10.8	
Prismatic Coefficient	0.68	
Auxiliary Horsepower	4-7	
Headroom	3'-6"	1.07 m

*__Caution__: The displacement quoted here is for the boat in coastal cruising trim. That is, with the fuel and water tanks filled, the crew on board, as well as the crews' gear and stores in the lockers. This should not be confused with the "shipping weight" often quoted as "displacement" by some manufacturers. This should be taken into account when comparing figures and ratios between this and other designs.

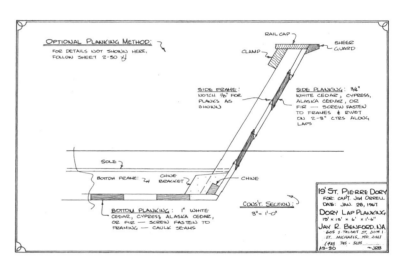

is by no means any less capable of taking her crew out and back in safety.

The Preliminary Study turned into the Profiles & Arrangements drawing, which I thought had all the information on the one sheet that a builder would need. Captain Jim Orrell thought we needed more drawings to make a realistically marketable plan set, so I created some more drawings, which are shown here too. The Cruising Version with its little cuddy is something I thought would be great fun and started to build it—but other life events got in the way of finishing it. I still think it's salty as all get out, though. . . .

Building Badger

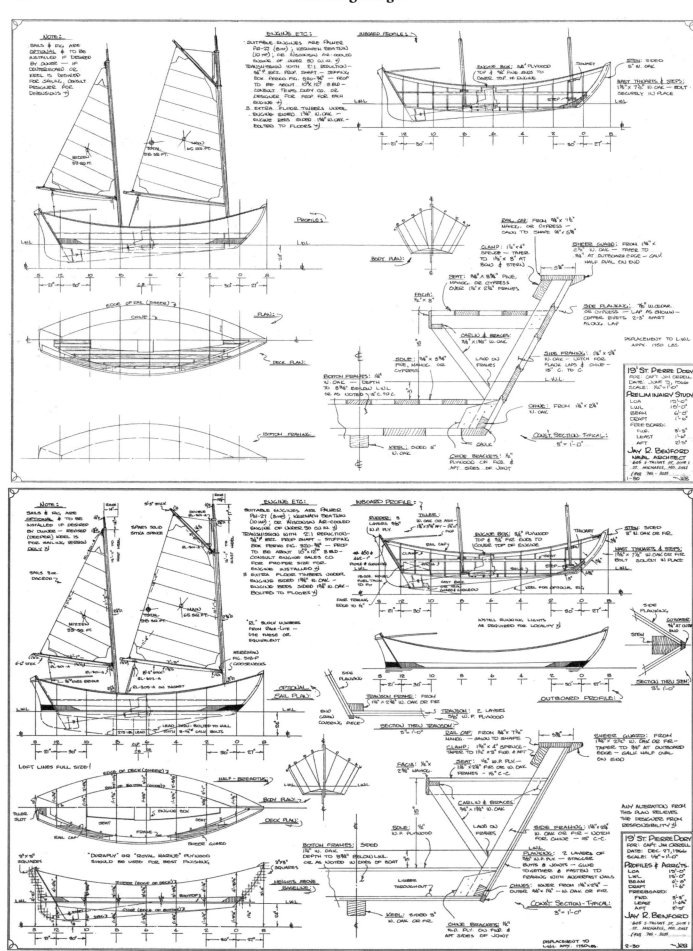

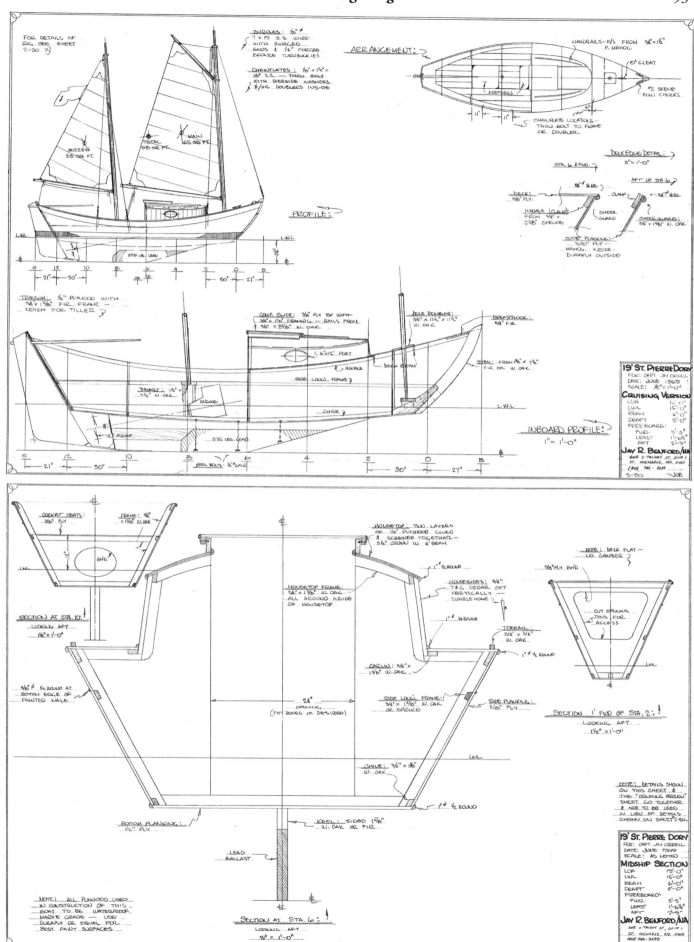

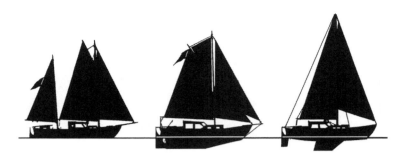

30' Sailing Dory
Design Number 32
1967

There are three primary versions of this design; the original gaff ketch version, the revised pilothouse sloop and the gaff rigged cutter version of the pilothouse sloop. The gaff ketch has a small aft cabin, providing separation for guests or older kids. For a voyaging version, I would suggest that the interior of the Marconi sloop, with a chart table like the gaff sloop in place of the inside helm, would be the best choice. It is a slightly smaller version of what is in the 34' **Badger** and has similar practical separation of the spaces.

The ketch shows the lead ballast fin keel. The Marconi sloop shows the concrete and scrap metal ballasted fin keel, with three inches of additional draft to keep the stability the same. This is a modified (fatter) fin keel that uses scrap metal inside a ferro-cement fin. The gaff sloop has a long keel and either lead or concrete and scrap metal for ballast. Again, the draft is increased with the concrete keel. Not shown, but possible to create, is a reduced draft long keel with lead on it which would keep the center of gravity of the keel where it should be for good sailing stability.

The indicated displacement is in coastal cruising use. For voyaging, I would assume that she would get loaded down a fair bit, like **Badger**, with the stores and supplies needed for such service. Fortunately, these dories take this loading gracefully, not unlike their predecessors the working dories which set out light and returned carrying tons of fish.

The raised deck versions, designed with both a tombstone transom (30'-5") and the double-ender (31'-8") have junk rigs, single or double masted, available. While they might be thought of as a **Baby Badger**, they are capable boats in their own rights, whilst not having as much carrying capacity as the 34-footer does. Still, they would make good cruising and liveaboard yachts. For more detail on the raised deck versions, see their chapter later in this book.

This simple-to-construct, no-nonsense little 30-footer will perform well. Though flat-bottomed in dory fashion, when she's heeled over, she presents a "v" to the water, and will move along at quite a good clip. She has 2,400 pound lead ballast on her keel, with plywood, hard chine construction over sawn frames. She has enclosed head, functional galley & roomy area for dining and lounging.

The 30 and 32' dories share rigs, keels and basic scantlings. The 4'-0" draft versions have lead ballast. The jibs on the ketches are on roller furling gear, so no bowsprit work will be needed.

The booms should have topping lifts fitted, and lazy jacks for ease in dousing sail. This kind of gear makes for much easier sailhandling and is favored by singlehanders.

Building *Badger*

The rudders on all the dories are oversized, so that they will be effective at low angles. Large rudder angles lead to increased drag, slowing the boats down, and can lead to a broach if the rudder loses its lift and stalls, causing the boat to loose steerageway. It will be easy to fit a trim tab on any of these rudders for self-steering gear.

If you look at the rudder in the original versions of this design, you will see an unbalanced blade, with the center of area well aft of the stern. Early feedback from builders suggested this required quite a bit of effort on the tiller. We then modified the rudder to put some "balance" to it, adding some area forward of the line of the pintles and gudgeons, to counteract the area aft. This significantly reduces the effort to steer her and I'd recommend doing one of these revised rudders when building her.

In looking at the evolution from the 1967 preliminary version of this design and the 1975 completed version, the hull form has been modified. The forward half of the boat is finer in plan view and the after part is fuller, changes that were done with an eye to making her a better performer under sail. The arrangement plan has also changed too, with what I think will be a useful cruising layout.

The pilothouse version was the second variation created, and this one has all the accommodations in the forward cabins. There is an inside steering station, though some folks have opted to turn this space into a built-in chart table.

The gaff topsail cutter rig was the third variation done. It shows the chart table instead of inside steering and storage in the forward end of the boat instead of double berth. The ballast is concrete and scrap metal instead of cast lead on the original version.

Table of Particulars:

30' Sailing Dory	Imperial	Metric
Length Overall	30'-0"	9.14 m
Length Datum Waterline	26'-0"	7.92 m
Beam	10'-0"	3.05 m
Draft	4'-0"	1.22 m
Draft—alternate keel	4'-3"	1.30 m
Freeboard: Forward	4'-6"	1.37 m
Freeboard: Least	2'-8¾"	0.83 m
Freeboard: Aft	3'-3½"	1.00 m
Displacement, Cruising Trim*	6,700	3,039 kg
Displacement-Length Ratio	170	
Ballast	2,680	1,216 kg
Ballast Ratio	40%	
Sail Area, Square Feet	500	46.45 sq m
Sail Area-Displacement Ratio	22.5	
Prismatic Coefficient	.58	
Pounds Per Inch Immersion	685	
Auxiliary Horsepower	9	
Water, Gallons	25	95 liters
Fuel, Gallons	25	95 liters
Headroom	6'-1"	1.85 m

*__Caution__: The displacement quoted here is for the boat in coastal cruising trim. That is, with the fuel and water tanks filled, the crew on board, as well as the crews' gear and stores in the lockers. This should not be confused with the "shipping weight" often quoted as "displacement" by some manufacturers. This should be taken into account when comparing figures and ratios between this and other designs.

Above—note steps and/or handholds built into the rudder. Right—structure complete and preparing for painting.

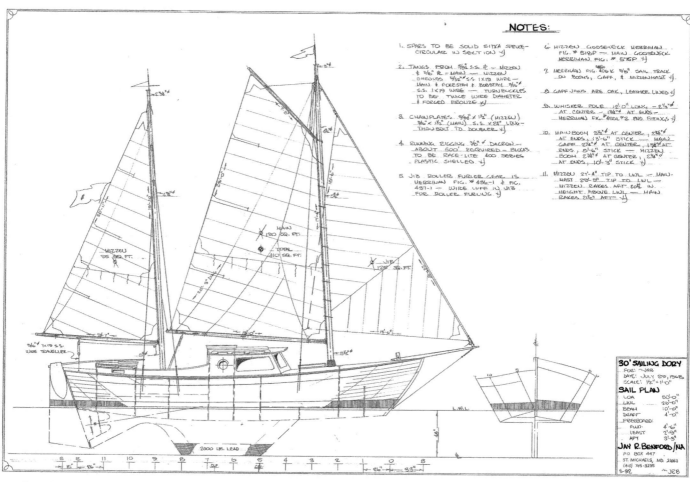

These drawings are the preliminary, or early, version of this design.

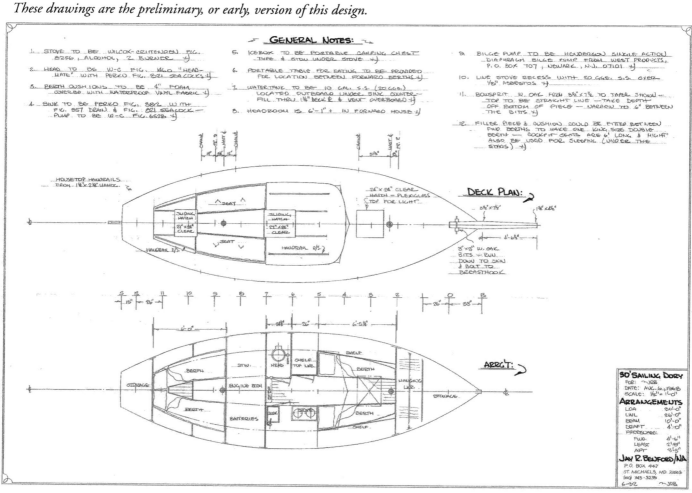

Building *Badger*

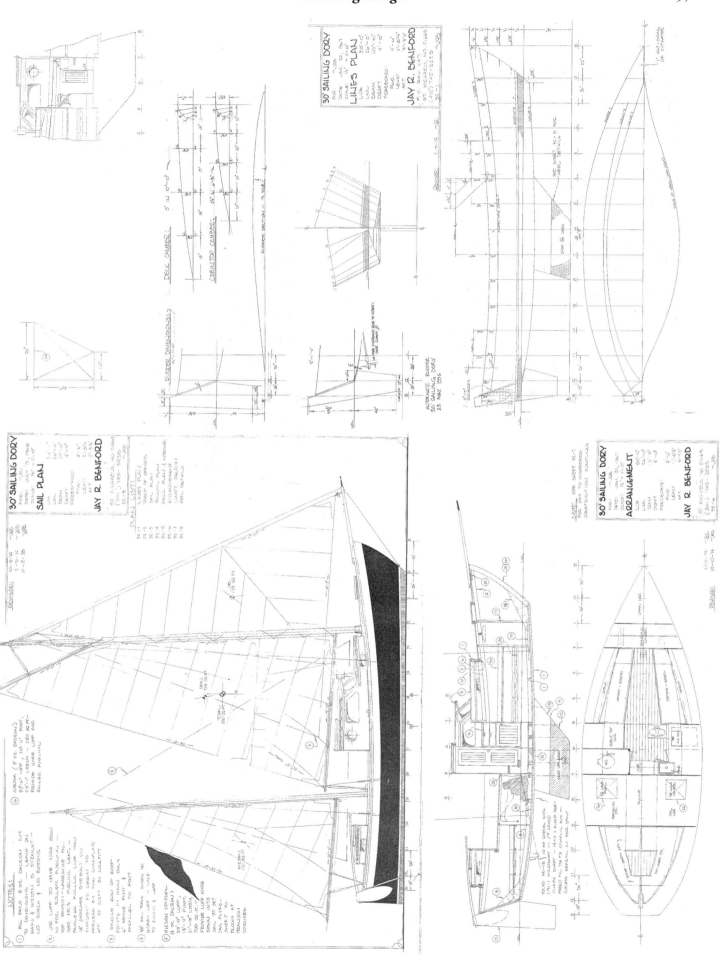

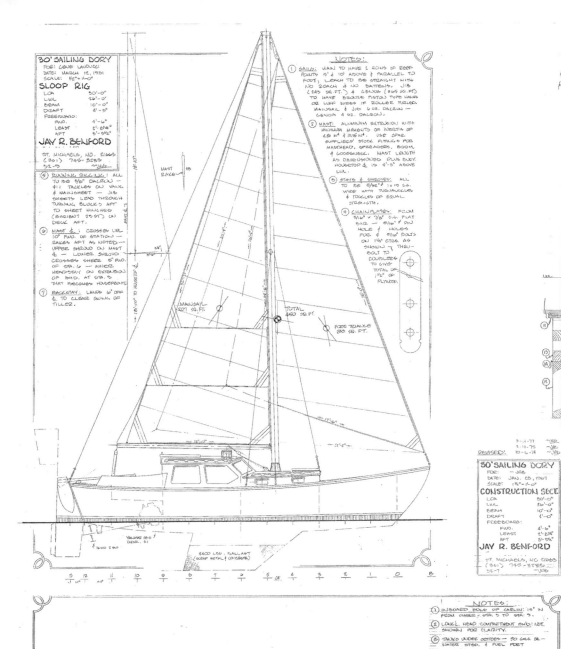
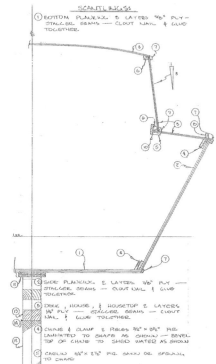
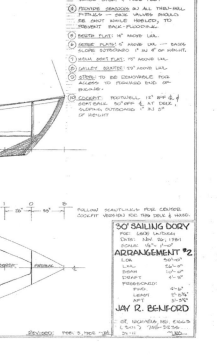

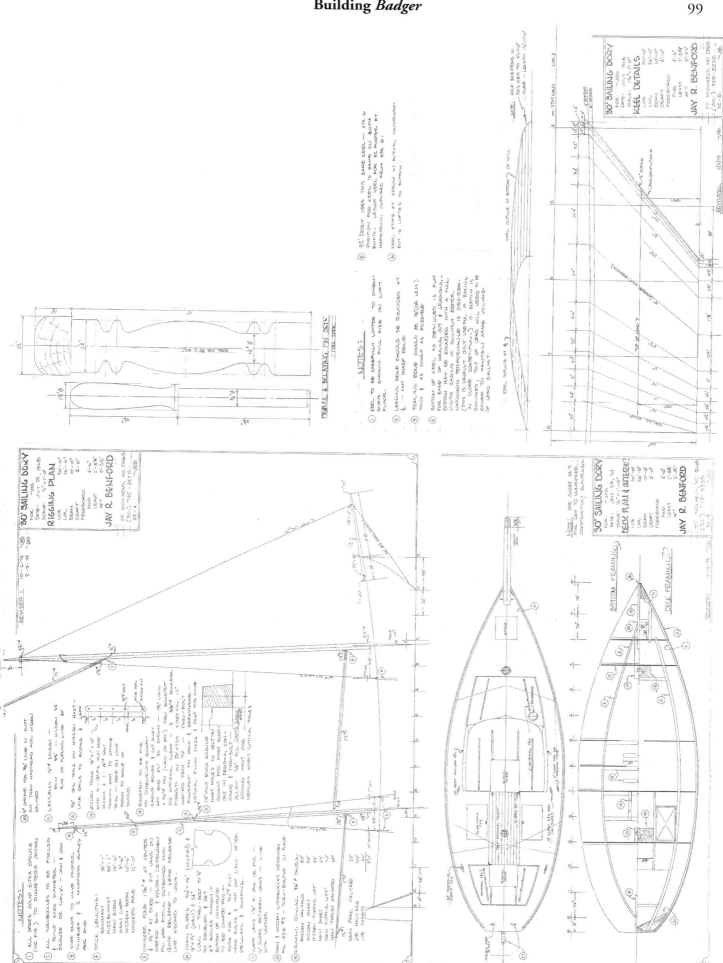

This sheet is a combination of inboard profile, deck plan, rigging plan and details on how to build the rig and pieces of it. Note the swept back spreaders and the single lowers that are also kept well aft. These will work together to keep the masthead where it should be and mean that, at least in moderate to light conditions, that the use of the running backstays won't always be needed. This is one of those things you learn in living with the boat and keeping an eye on her as you're enjoying sailing and cruising in her. More on this version is on the following page.

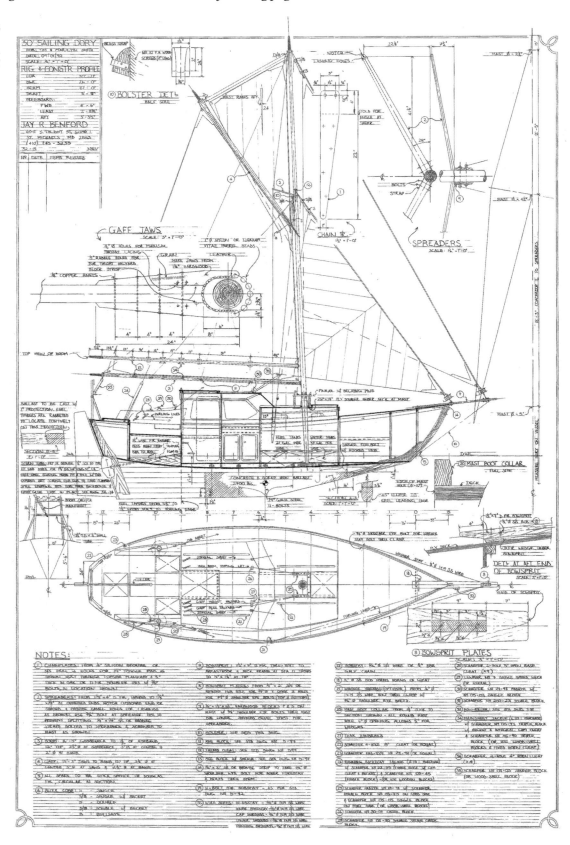

Building Badger

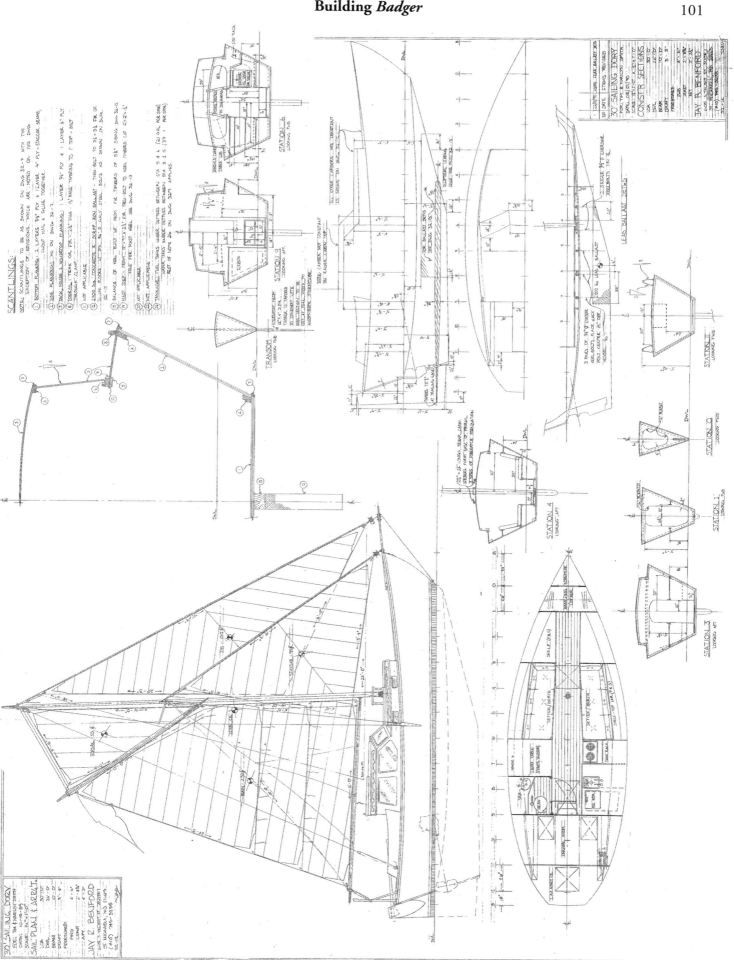

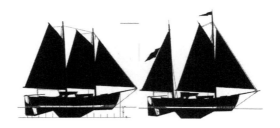

27' DE Ketch or Schooner
Design Number 33
1966 & 1969

This design was developed as an idea for a more burdensome 27' Sailing Dory that could carry more people and stores. In looking at the two versions—the original schooner idea and the later ketch version—the ketch has more appeal to me, visually. The schooner suffers from, perhaps, too short a rig. This is one lesson that a yacht designer needs to learn early on—don't let the size of the sheet of drawing paper dictate the design. Or—plan ahead better when you're working on placing the view of the boat so that there is adequate room for all that you would like to see included on that drawing. As

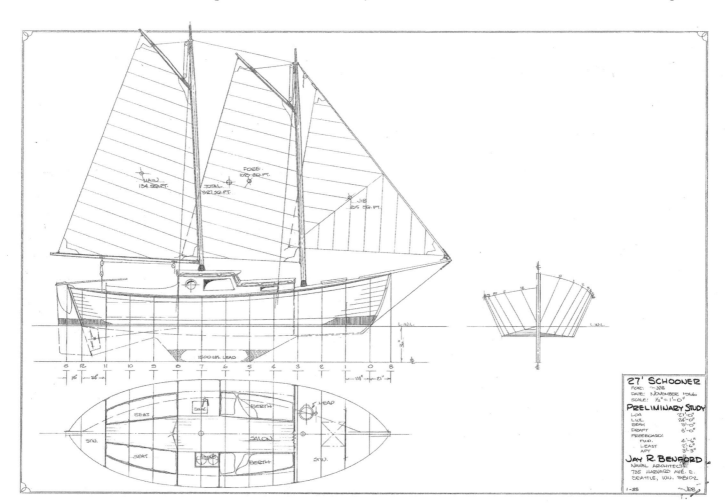

Building *Badger*

Table of Particulars:

27' DE Ketch or Schooner	Imperial	Metric
Length Overall	27'-0"	8.23 m
Length Datum Waterline	24'-0"	7.32 m
Beam	9'-0"	2.74 m
Draft	3'-0"	0.91 m
Freeboard, Forward	4'-6"	1.37 m
Freeboard, Least	2'-6"	0.76 m
Freeboard, Aft	3'-3"	0.99 m
Sail Area—Schooner	327 sq. ft	30.38 m²
Sail Area—Ketch	350 sq. ft.	32.52 m²
Headroom	5'-3"	1.60 m

*Caution: The displacement quoted here is for the boat in coastal cruising trim. That is, with the fuel and water tanks filled, the crew on board, as well as the crews' gear and stores in the lockers. This should not be confused with the "shipping weight" often quoted as "displacement" by some manufacturers. This should be taken into account when comparing figures and ratios between this and other designs.

one gains experience in preliminary design work, these issues are more skillfully handled. If you look at the ketch version, done a three years later, you will see that her rig looks more in proportion to the boat.

Today, I think that I'd work up a new and improved interior layout for cruising with a couple aboard. Perhaps it's my changing life, but I think it would be nice to have a single, wide berth instead of the well separated settee berths. And an enclosed head for the ladies to have some privacy. With a little diesel and a nice galley, she'd be a great way to go cruising.

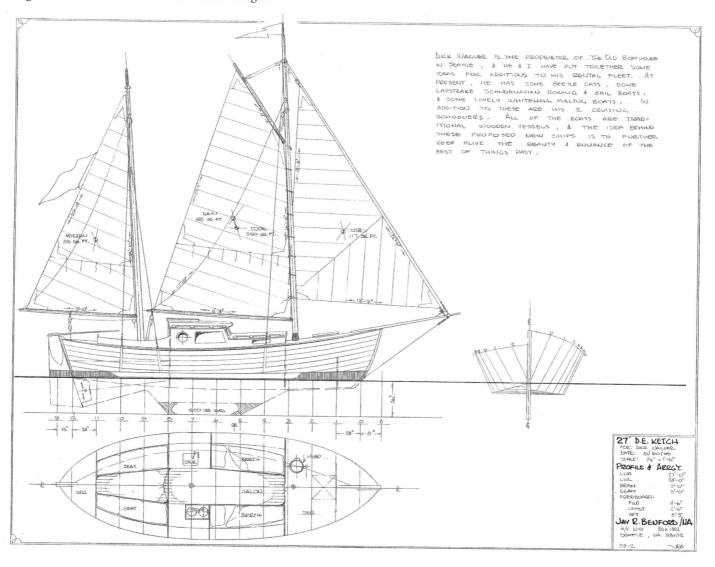

32' Sailing Dory *Shoestring*
Design Number 36
1968

The first 32-footer was built and named **Shoestring** by her builders. We so liked the name—and the thought that she could be built on the proverbial shoestring—that we adopted the name for the design. As the narrowest of these dory designs, this boat is the most sensitive to weight aloft, and care needs to be taken that her stability is not compromised, by putting too heavy a rig on her. Sail carrying power, and thus the ability to sail out of a close situation, is directly related to the stability of the boat. Anything that can be done to minimize or reduce weight aloft will aid the stability. All things being equal, transverse stability is directly related to waterline beam cubed. Thus, a beamier boat will have better ability to carry a heavier rig. An aluminum spar version of the 30-footer's sloop rig, having the least weight aloft, would be the best choice to give this boat the most stability.

A simple change in filling the berth flat all the way across the forward cabin will yield a nice double berth there. I think this would be an improvement, particularly with a sloop rig that would move the mast back to the aft end of the dining table and thus not breakup the head of the new double berth.

The rudders on all the dories are oversized, so that they will be effective at low angles. Large rudder angles lead to increased drag, slowing the boats down, and can lead to a broach if the rudder loses its lift and stalls, causing the boat to loose steerageway. It will be easy to fit a trim tab on any of these rudders for self-steering gear.

Table of Particulars:

32' Sailing Dory	Imperial	Metric
Length Overall	32'-0"	9.75 m
Length Datum Waterline	27'-0"	8.23 m
Beam	9'-0"	2.74 m
Draft	4'-0"	1.22 m
Freeboard: Forward	4'-2"	1.27 m
Freeboard: Least	2'-6¼"	0.77 m
Freeboard: Aft	3'-6½"	1.08 m
Displacement, Cruising Trim*	6,900	3,130 kg
Displacement-Length Ratio	156	
Ballast	2,760	1,252 kg
Ballast Ratio	40%	
Sail Area, Square Feet	500	46.45 sq m
Sail Area-Displacement Ratio	22.1	
Prismatic Coefficient	.60	
Auxiliary Horsepower	9	
Water, Gallons	25	95 liters
Fuel, Gallons	25	95 liters
Headroom	6'-1"	1.85 m

*****Caution**: The displacement quoted here is for the boat in coastal cruising trim. That is, with the fuel and water tanks filled, the crew on board, as well as the crews' gear and stores in the lockers. This should not be confused with the "shipping weight" often quoted as "displacement" by some manufacturers. This should be taken into account when comparing figures and ratios between this and other designs.

Shown below is the preliminary version of this design.

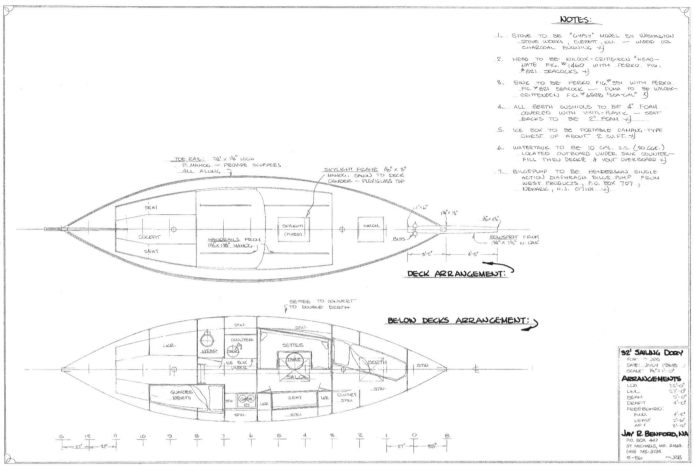

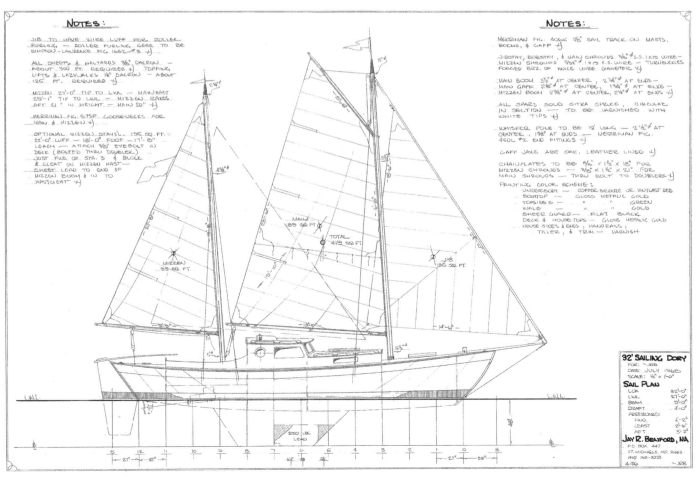

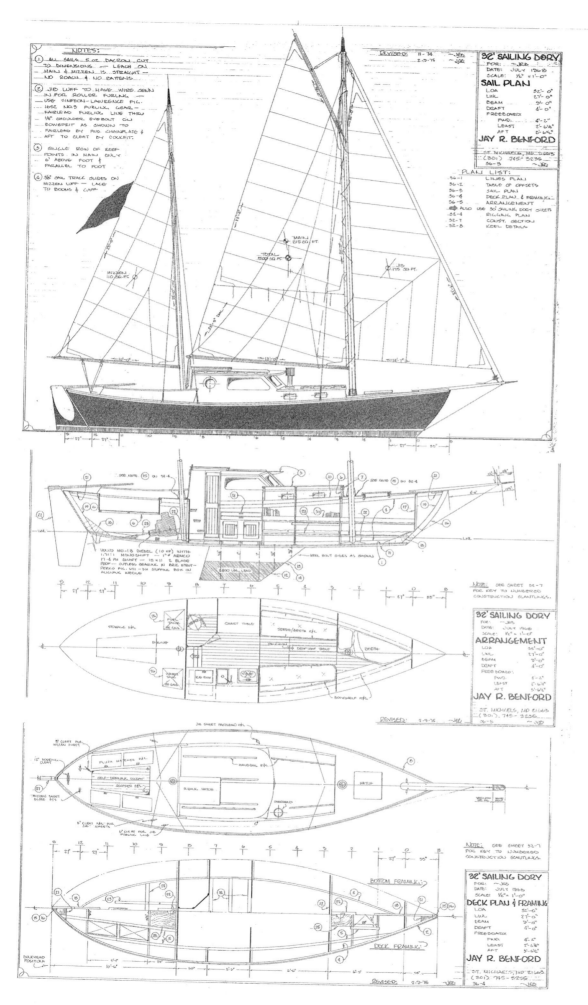

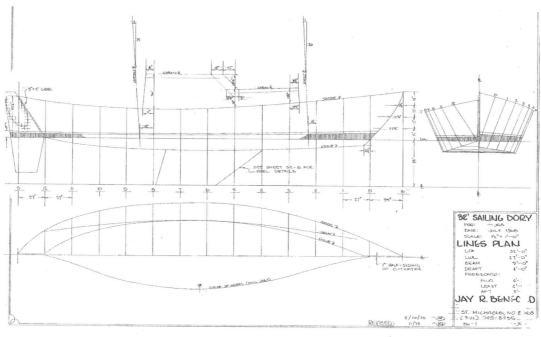

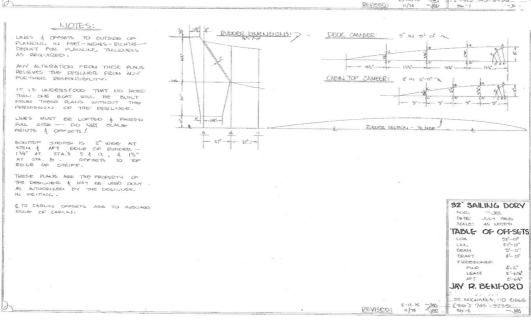

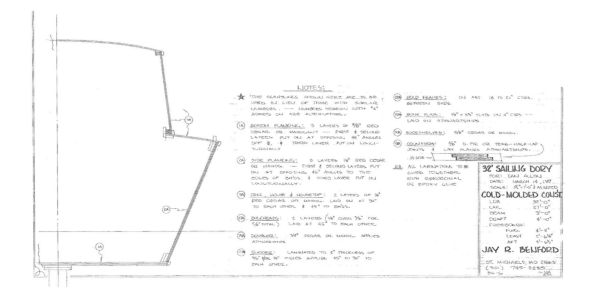

26' V-Bottom Ketch
Design Number 41
1975

Table of Particulars:

26' V-Bottom Ketch	Imperial	Metric
Length Overall	26'-0"	7.92 m
Length Datum Waterline	24'-0"	7.32 m
Beam	7'-0"	2.13 m
Draft—fin keel	3'-6"	1.07 m
Freeboard, Forward	3'-10"	1.17 m
Freeboard, Least	1'-10"	0.56 m
Freeboard, Aft	2'-6"	0.76 m
Displacement, Cruising Trim*	7,150 lbs.	3,243 kg
Displacement-Length Ratio	231	
Ballast	2,000	907 kg
Ballast Ratio	28%	
Sail Area, Square Feet (m2)	281/350	26.11/32.52
Sail Area-Displacement Ratio	12.11/15.09	
Auxiliary Horsepower	5 to 8	
Water, Gallons	20	76 litres
Fuel, Gallons	10	38 litres
Headroom	4'-6" - 5'-3"	1.37-1.6 m

*****Caution**: The displacement quoted here is for the boat in coastal cruising trim. That is, with the fuel and water tanks filled, the crew on board, as well as the crews' gear and stores in the lockers. This should not be confused with the "shipping weight" often quoted as "displacement" by some manufacturers. This should be taken into account when comparing figures and ratios between this and other designs. Also, loading down for voyaging and living aboard, as with the Hill's ***Badger***, will add considerably to these figures, perhaps as much as 50%. Designs like the 34' ***Badger***, which load down gracefully and still sail well, make a good choice for anyone wanting to go voyaging on a small income.

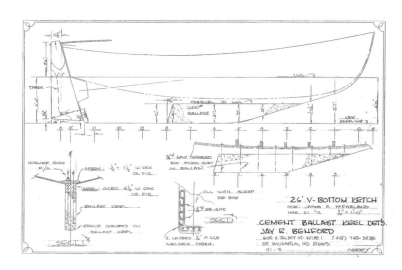

This design was developed as a v-bottom version of the 26' Sailing Dory. Comfortably laid out for three, she is intended for coastwise cruising. She has a lot of simple detailing, like her transom stern, outboard rudder and optional cement keel. A small inboard could be fitted. There are two rigs that were created, one for light conditions and one for heavy wind areas. Of course, one could build the larger rig and be prepared to reef when the wind is stronger, giving a wider range of cruising areas in which the boat would be well suited to be cruising.

Building *Badger*

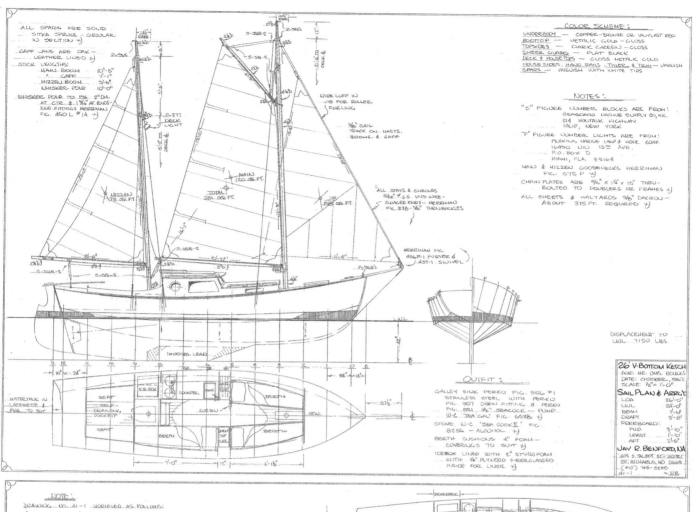

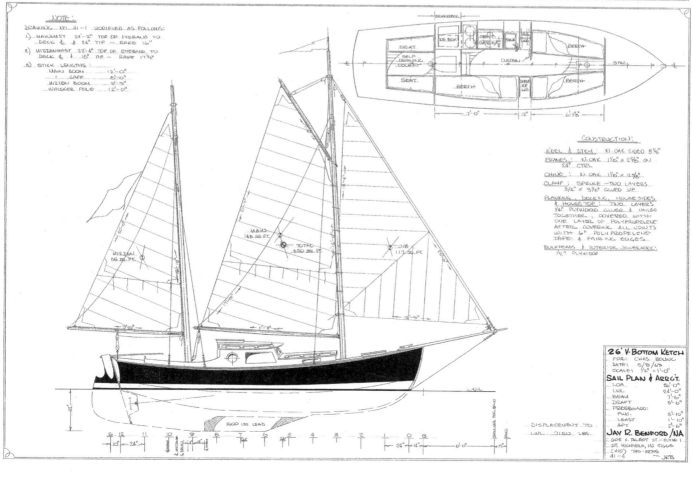

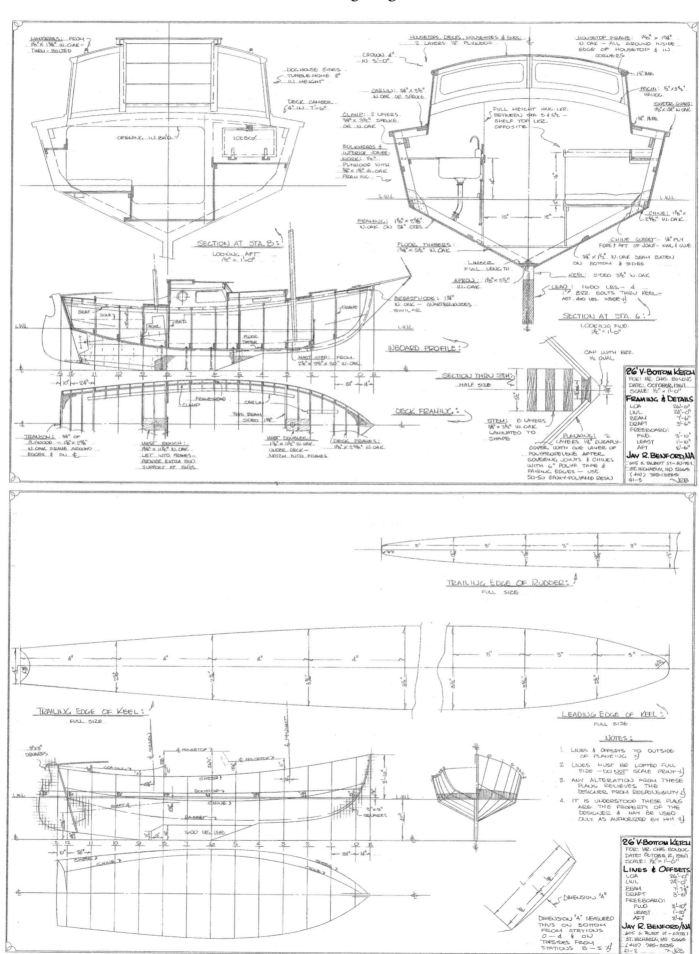

32' St. Pierre
Designs Number 113
1974

The 32' St. Pierre Dory *Proctor* is designed to be built in plywood. She is a traditional St. Pierre Power Dory of the type that has proven a practical fishing vessel. She has comfortable living accommodations for her size, a good fish hold, and a working trolling cockpit. Her ability to get out to the fishing grounds quickly yet with reasonable comfort, and to keep to sea in a variety of conditions makes her an attractive investment. Her plans use commonly available lumber yard materials, making her a practical choice for a simple working boat to be built on a budget.

She was named *Proctor* by our original client for her. Admiralty attorneys are sometimes referred to as Proctors and, if I recall correctly, he was one and the name seemed fitting. In the years after building her, she was used successfully as a salmon troller in the Pacific, off the Oregon coast.

It would be possible to work out other versions, without the fish hold, and with expanded cruising accommodations.

In the course of designing *Proctor*, we had some computer work done to check on her stability. The following two sets of curves are a summary of that data. One shows the results with her floating at her designed waterline and the other set of curves shows her loaded down an additional six inches with cargo. You can see how her stability changes with the loading and how she still has good righting arms.

Table of Particulars:

32' Power Dory	Imperial	Metric
Length Overall	32'-0"	9.75 m
Length Datum Waterline	25'-6"	7.77 m
Beam	10'-10½"	3.31 m
Draft	2'-8"	0.81 m
Freeboard, Forward	4'-8½"	1.44 m
Freeboard, Least	2'-0¼"	0.62 m
Freeboard, Aft	4'-2½"	1.28 m
Displacement, Cruising Trim*	8,660 lbs.	3,928 kg
Displacement-Length Ratio	233	
Prismatic Coefficient		
Pounds Per Inch Immersion	742	
Auxiliary Horsepower		
Water, Gallons	100 Gals.	379 litres
Fuel, Gallons	100 Gals.	379 litres
Headroom	5'-7" to 6'-6"	1.78-1.98 m

***Caution**: The displacement quoted here is for the boat in coastal cruising trim. That is, with the fuel and water tanks filled, the crew on board, as well as the crews' gear and stores in the lockers. This should not be confused with the "shipping weight" often quoted as "displacement" by some manufacturers. This should be taken into account when comparing figures and ratios between this and other designs.

To get the righting moment, the actual working stability of the dory, multiply the righting arm by the displacement and get the righting moment.

A decade later, about 1984, we finally made the investment in a computer design system that let us do all this analysis in-house, as part of the initial design process. The new computer system also let us provide computer faired lines and full-sized patterns. And take quick looks at the stability of the boat while we're still refining the hull form.

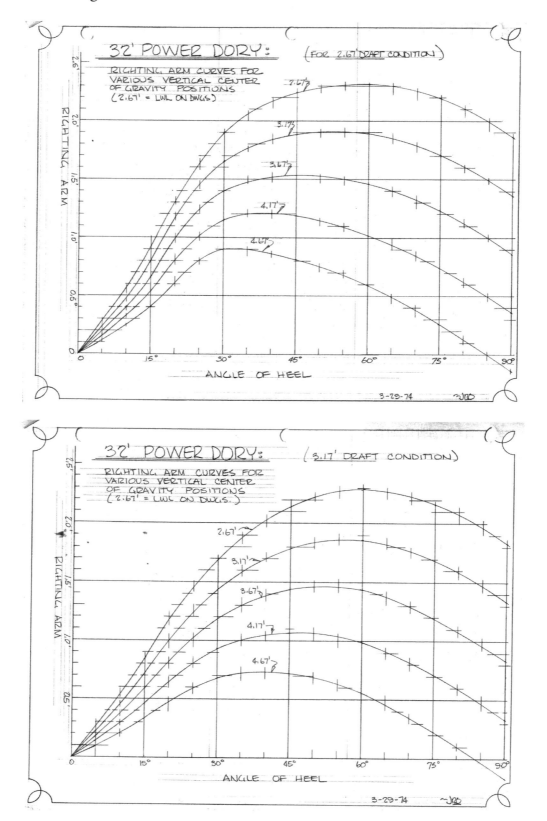

Building *Badger*

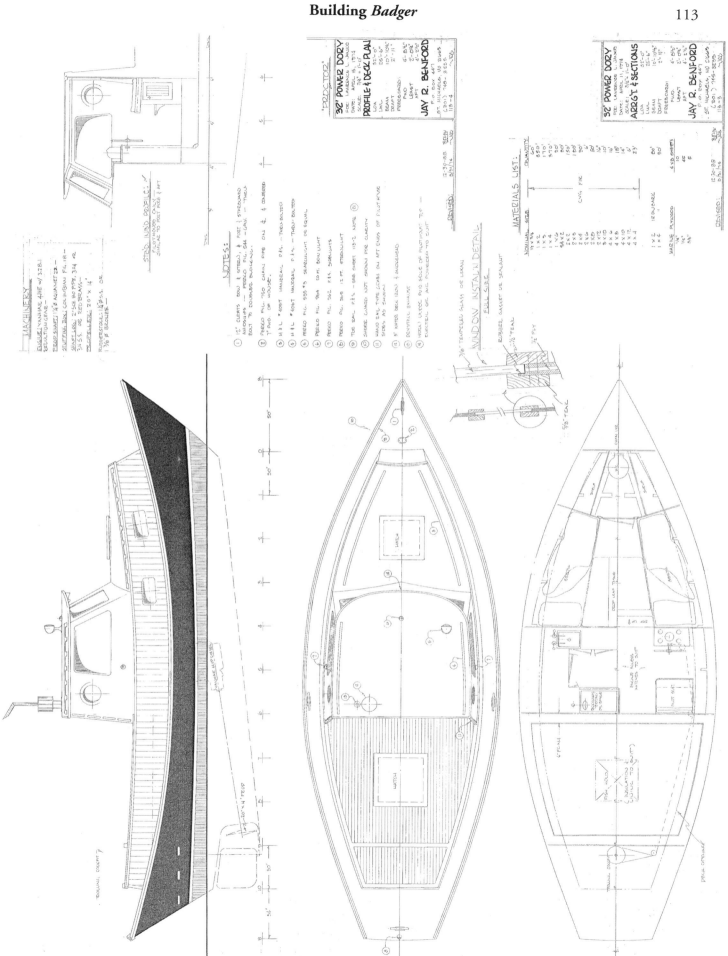

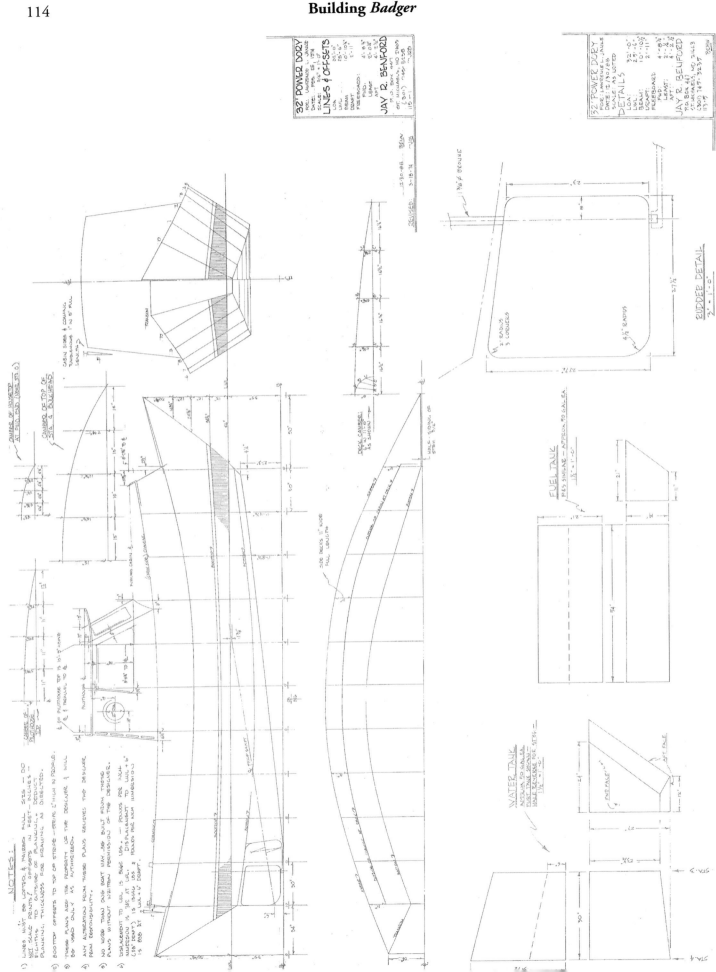

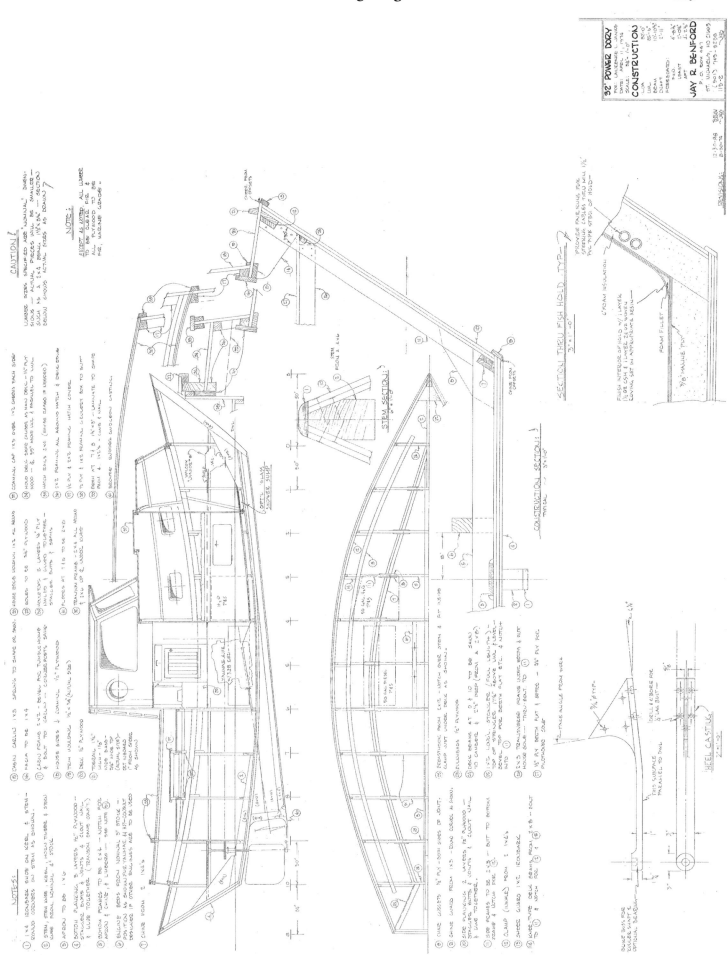

36' Sailing Dory *Donna*
Design Number 127
1975

Table of Particulars:

36' Sailing Dory	Imperial	Metric
Length Overall	36'-0"	10.97 m
Length Datum Waterline	31'-0"	9.45 m
Beam	11'-0"	3.35 m
Draft—fin keel	4'-6"	1.37 m
Draft—long, shallow keel (opt.)	3'-6"	1.07 m
Freeboard, Forward	5'-0¾"	1.54 m
Freeboard, Least	3'-0¼"	0.92 m
Freeboard, Aft	4'-2¾"	1.29 m
Displacement, Cruising Trim*	13,425	6,008 kg
Displacement-Length Ratio	201	
Ballast	5,350	2,247 kg
Ballast Ratio	40%	
Sail Area, Square Feet	700	65.03 m²
Sail Area-Displacement Ratio	19.83	
Prismatic Coefficient	.63	
Pounds Per Inch Immersion	983	
Auxiliary Horsepower	27	
Water, Gallons	110	416 litres
Fuel, Gallons	55	208 litres
Headroom	6'-4"	1.93 m

*Caution: The displacement quoted here is for the boat in coastal cruising trim. That is, with the fuel and water tanks filled, the crew on board, as well as the crews' gear and stores in the lockers. This should not be confused with the "shipping weight" often quoted as "displacement" by some manufacturers. This should be taken into account when comparing figures and ratios between this and other designs. Also, loading down for voyaging and living aboard, as with the Hill's *Badger*, will add considerably to these figures, perhaps as much as 50%. Designs like the 34' *Badger*, which load down gracefully and still sail well, make a good choice for anyone wanting to go voyaging on a small income.

This design has the same nice interior space layout as *Badger*, with the addition of quarter berths aft. For a family of four to live aboard, as has been done on one of the sisterships, this provides good separation of the sleeping spaces. I would put a double berth in the forward cabin, like *Badger*'s, and as shown on the 37½, and have the saloon berths and quarter berths for sleeping in on a passage. This interior and trunk cabin can be used on the 37½' cutter, if an alternate version is wanted. The trunk cabin version of the 37½' dory shows double-berth-forward accommodations variation that I would recommend for a couple or family wanting to live aboard.

Several of these 36-footers have done extensive ocean voyaging and the owners report that they have been very comfortable. They also report that the boats are good at making quick passages and will surf in some trade wind conditions.

This relatively light displacement sailing dory with her 5,000 pounds of ballasted keel will move along at a most respectable pace, and still maintain a good deal of stability and solid comfort for her crew. Designed for offshore cruising and simplicity of construction, this practical vessel has very livable accommodations. Quarter berths aft will often be the most comfortable sleeping place in a seaway. The head with shower and oilskin locker is conveniently close to the cockpit, and U-shaped galley to starboard. A large chart table is next to the head, with settee/berths midships around drop leaf table. The foc's'le is home to V-berths and chain locker in the forepeak. A fair amount of stowage space can be found throughout the vessel, and especially in the large lazarette.

The gaff ketch rig was the original rig. The Marconi ketch was done after the first one had done some Pacific crossings. The owner thought having a taller rig would help catch the wind when the boat was down in the big ocean swells and keep her moving better. A cutter rig from the 37½' cutter could be adapted—or you could just build the 37½—and will probably give a bit better performance to windward. For most cruising folks, the enjoyable cruising is mostly done with the wind abeam or aft of the beam, so the slight difference to windward should not be the deciding factor in choosing a rig.

*Right: 36' **Donna** off Maui, October 1979, on one of her early voyages. She was built by Fred Schreiner in Alberta in only 13 months. She's since been to Mexico from Vancouver and to Hawaii and back, making many passages, proving herself a first rate sea boat.*

*These photos show Bruce Talbot's 36' **Windhover** under construction in Vancouver. At left, the side planking is on and the bottom planking follows next. Below the bottom is completed and she's been rolled over. Some time later, she was completed and became the family's home.*

The LINES & OFFSETS shows the 36 to have a somewhat fuller shape, which is reflected in her prismatic coefficient ratio number. Not that this should be thought of as an limiter to her performance. Several of her builders have reported to us experiencing her surfing in their open ocean cruising. Certainly that would be exciting!...

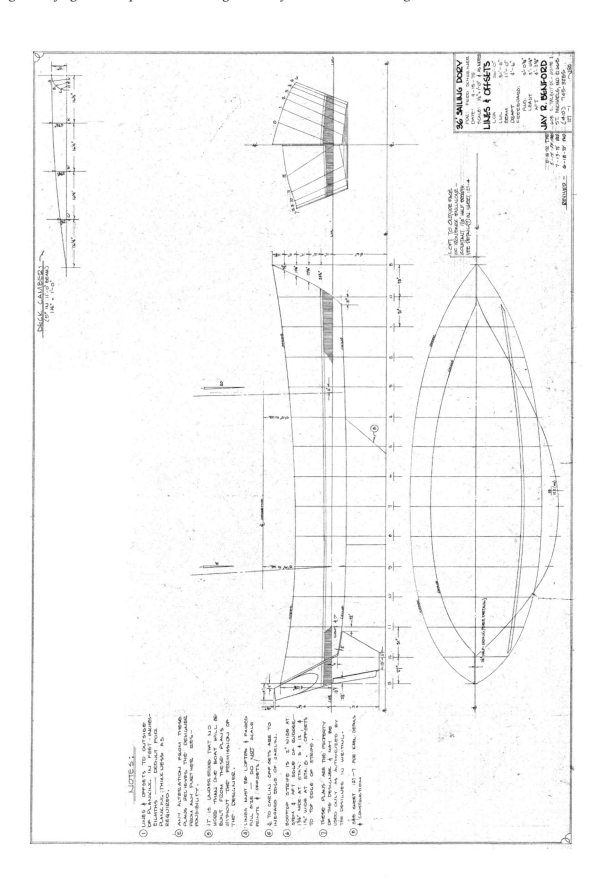

The original SAIL PLAN & ARRANGEMENT shows that she has nice separation of the sleeping quarters with fore and aft berths. The forward area could readily be converted to a double—this variation is shown on one of the 37½' versions. She has a Marconi mizzen, gaff main, and roller furling jib. Also noted in the sail dimensioning table are a mizzen staysail and a larger Genoa jib.

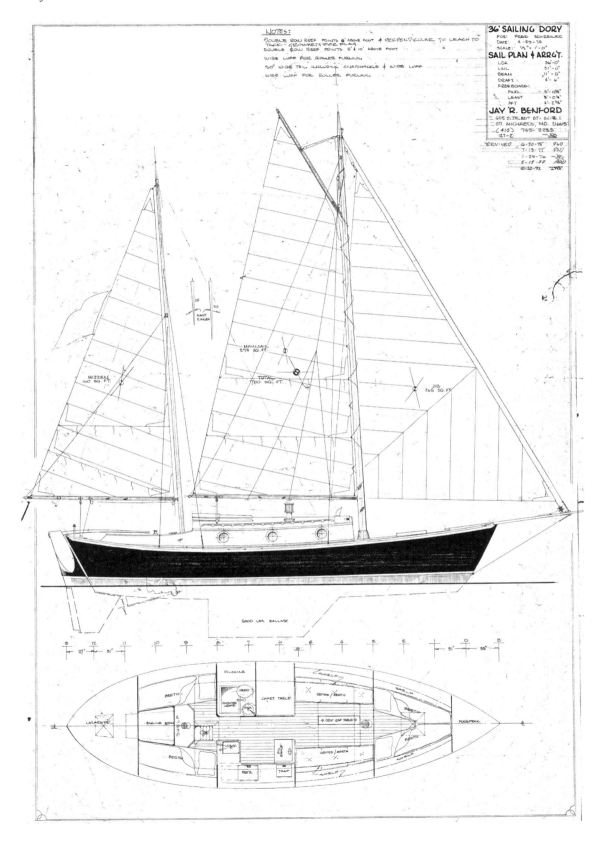

The INBOARD PROFILES & FRAMING gives a good look at both sides of her interior. These elevations are useful in understanding how the spaces are utilized. The plan view at the bottom of the sheet has the bottom framing on one side and the deck framing on the other. Since we assume that both sides will be built as mirror images of what is drawn, we don't usually draw both sides, unless there is to be some asymmetry in the structure.

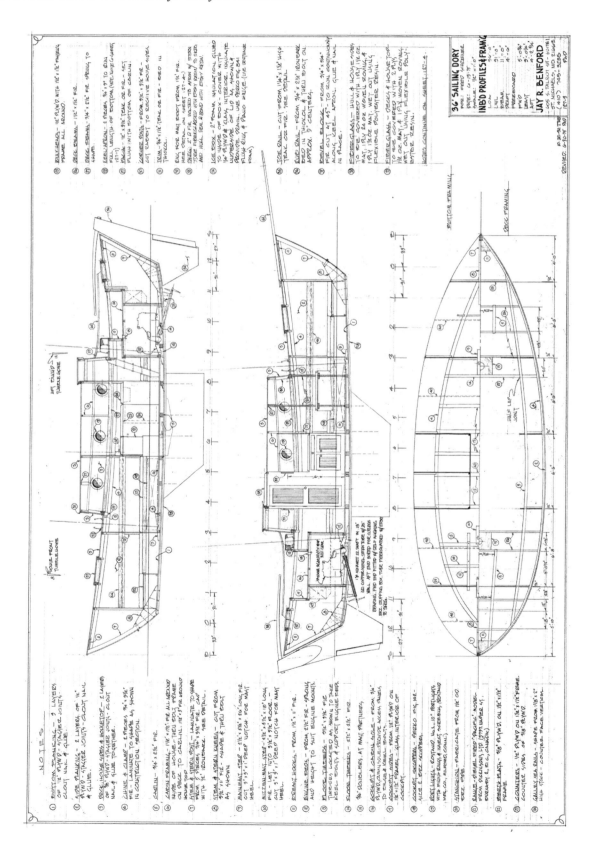

Building *Badger*

The SECTIONS & DETAILS sheet has a larger scale scantling section plus sections through many of the pieces of the structure and complete hull sections at several locations fore and aft. These are done to help define our expectations of how the boat is all expected to fit together.

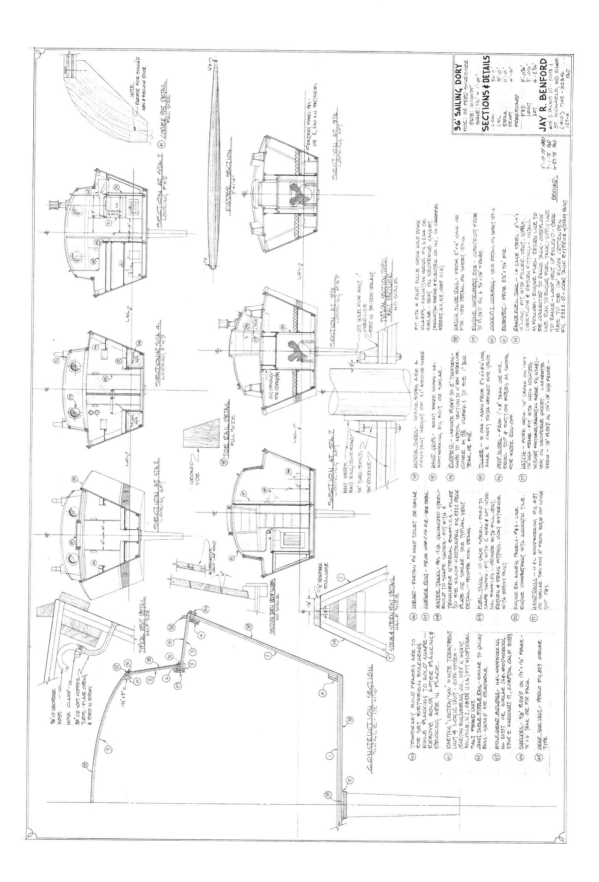

The RIGGING & DECK PLANS sheet has a lot of detail on it about how the rig is built and arranged on the boat. It also has a number of details showing how to build important parts of the rig.

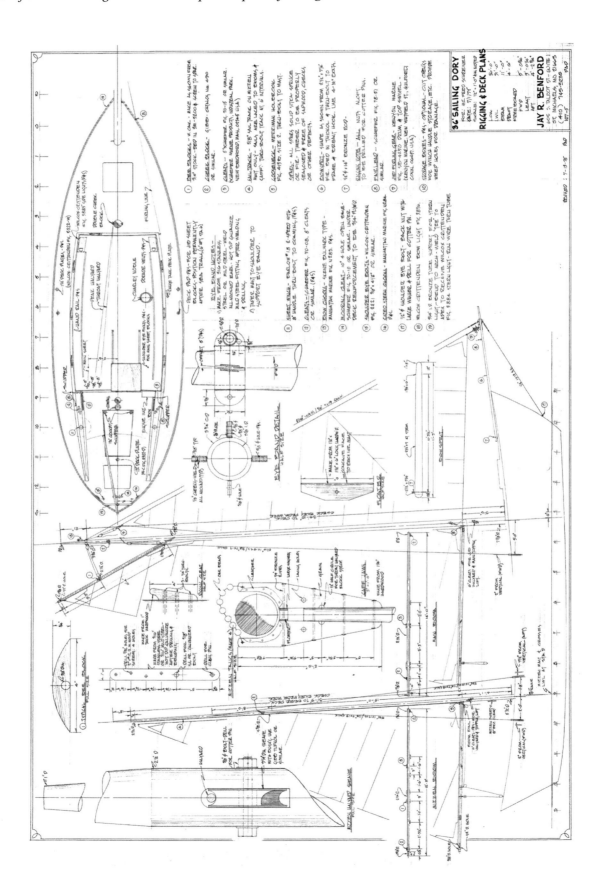

The RIGGING & DETAILS sheet has some additional details on building parts of the rig, the coaming, the companionway hatch, the chainplates, and fittings for the bowsprit. And, there is a table showing all the standing rigging and another one with all the running rigging on it.

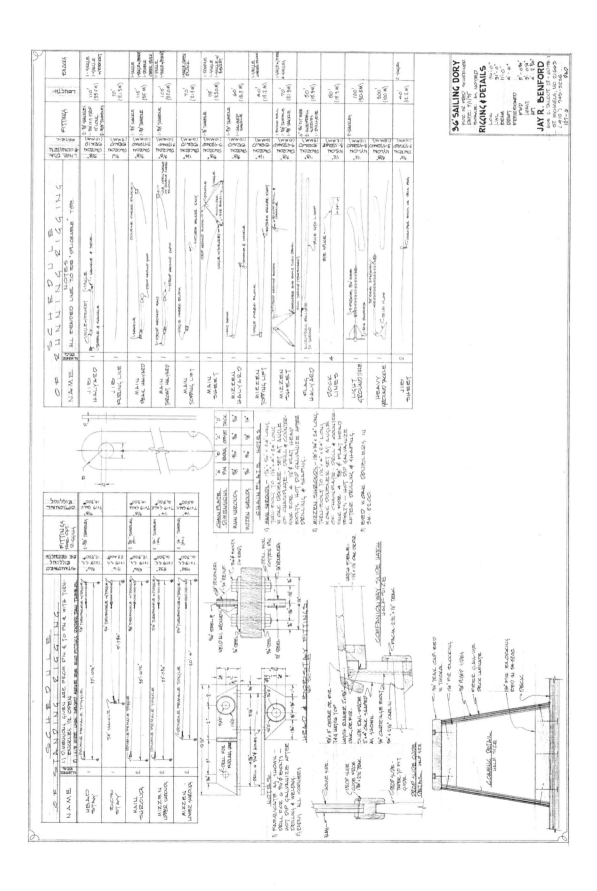

The KEEL DETAILS sheet shows her NACA foil section keel, built up of rebar welded together, covered in welded mesh, and filled with concrete and scrap metal to give the required weight and stability. The spacing of the keel bolts, welded in place, corresponds to the floor timbers that run from chine to chine. They keep the bottom from bending under the load of the keel as the boat heels.

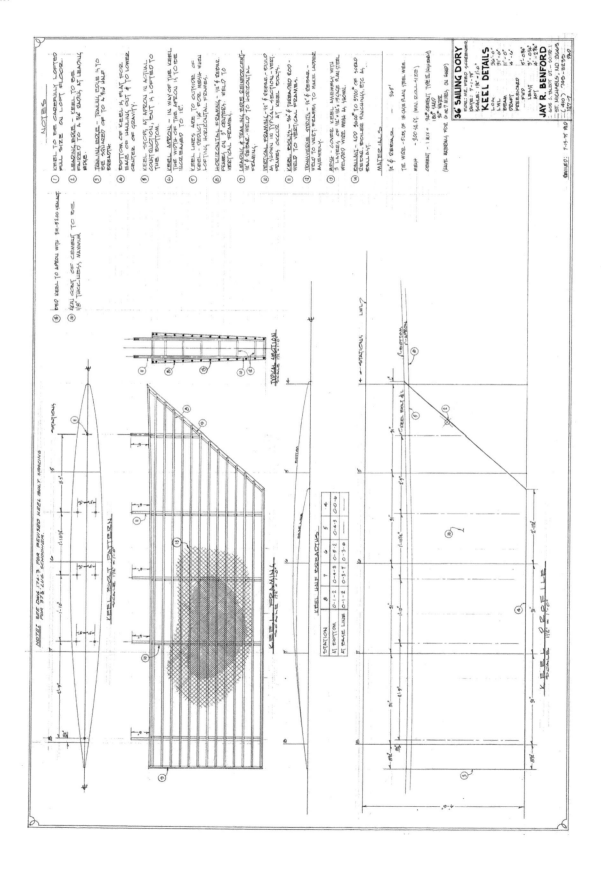

This sheet details the ALUMINUM CONSTRUCTION alternative for building the 36 in aluminum instead of plywood and epoxy. The multitude of transverse framing is what was called for in Lloyd's rules when we did the conversion.

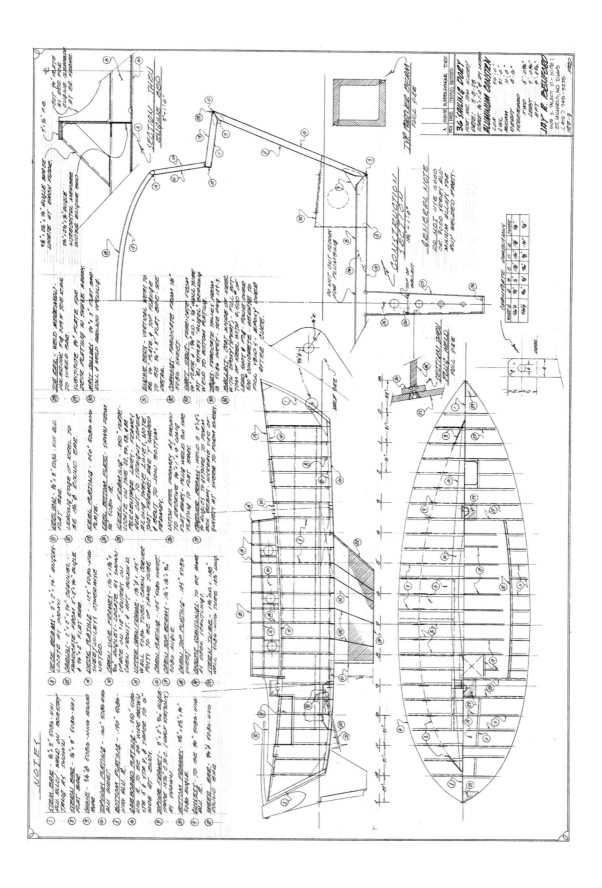

SAIL PLAN NO. 2 *After doing several Pacific Ocean crossings in* **Donna** *the builder came back to us asking for a bigger rig. He wanted something taller that would keep the sails filled while she was down in the trough between wave crests. We increased the sail area from 700 to 900 and gave her quite a bit more height to the masthead. This worked well for him and he seemed quite pleased with the result. And kept voyaging without making more changes to the rig.*

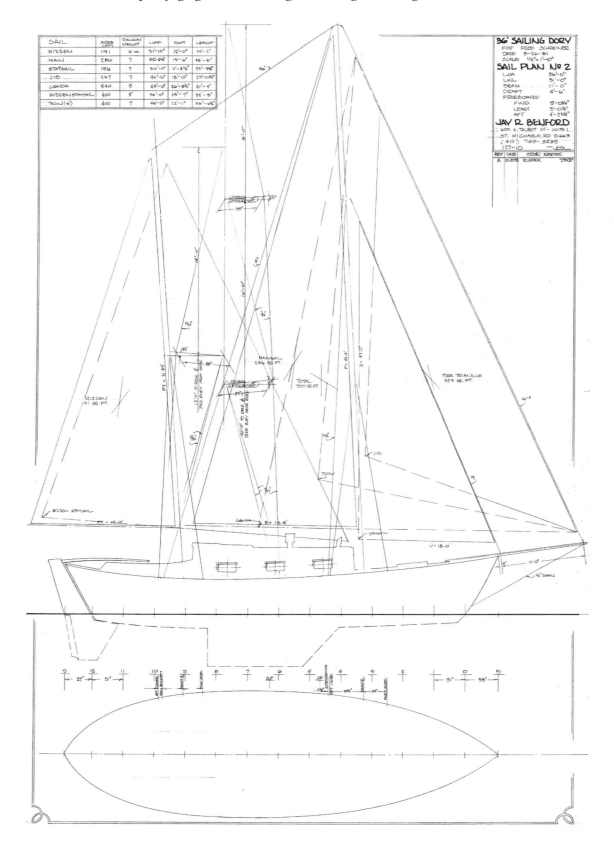

The OPTIONAL TALL RIG shows her appearance with the taller rig.

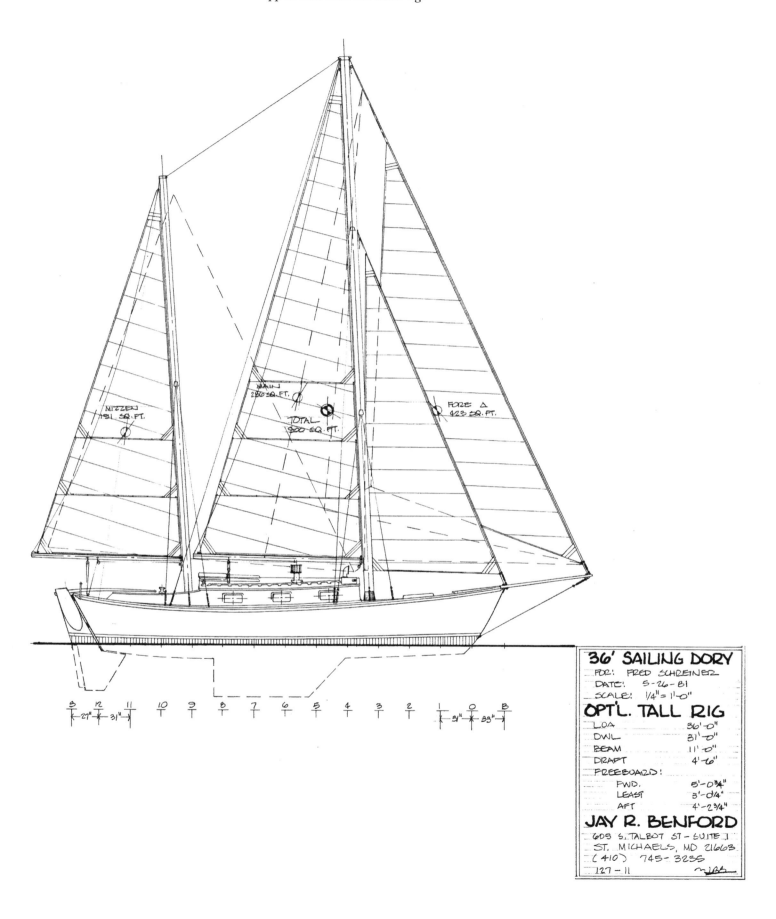

Left: **Donna** off Friday Harbor as she was brought by to show the finished results to the designer. Who fired up his **Sunrise** and went out as chase/photographers boat to record her performance.

Below: You can see that there are not yet whitecaps forming, yet **Donna** is charging right along at a great clip! We were all delighted to see how well she sailed. The promise she showed here was proven in the wonderful voyaging she did over many years.

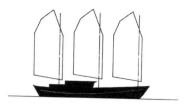

60' Aluminum Sharpie
Design number 140
1976

This is a concept for a boat that never got any further than the drawings that you see on this and the following page. The boat was to be built in welded aluminum, and we laid out a shape that could be readily achieved using large sheets of aluminum. The rig was created to get 45' bridge clearance, since this is a common height on ICW bridges that are not the higher 65' clearance ones. Using three sails to get sufficient sail area to drive her meant that we didn't have to go to a taller rig. Ever since showing the little silhouette of her rig (above) in **Voyaging On A Small Income**, we've had inquiries about the design, so we included her here so that all the drawings that exist—to date—can be seen.

Her planned layout, as defined by the bulkhead placement and notes on the Preliminary drawing indicate the spaces were, starting from the stern, a lazarette, stateroom, head, pilothouse over the engine room with hold forward of it, port and starboard settees, galley and head, staterooms, stateroom, forepeak. Her

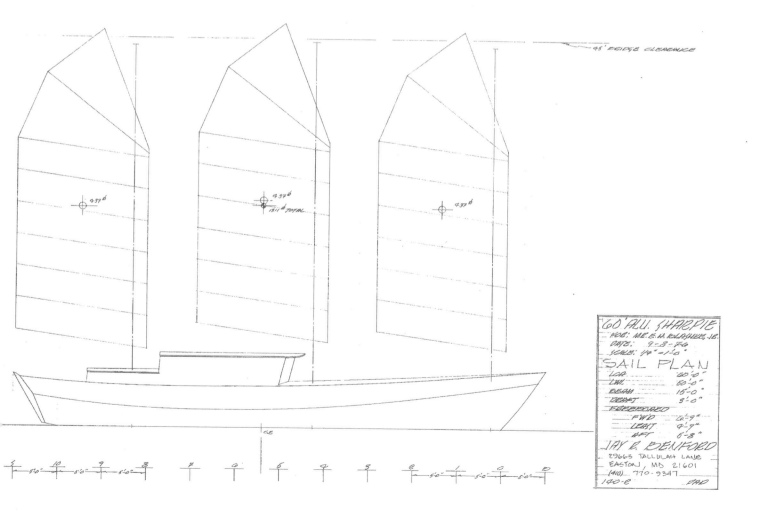

sail area is 1,311 square feet in three equal sails. The displacement is 42,000 pounds, giving a displacement-length ratio of 150 and sail area-displacement ratio of 17.35. her prismatic is 0.53 and her sail area-wetted surface is about 2.45. The notes in the file indicate we were looking at leeboards when the project halted.

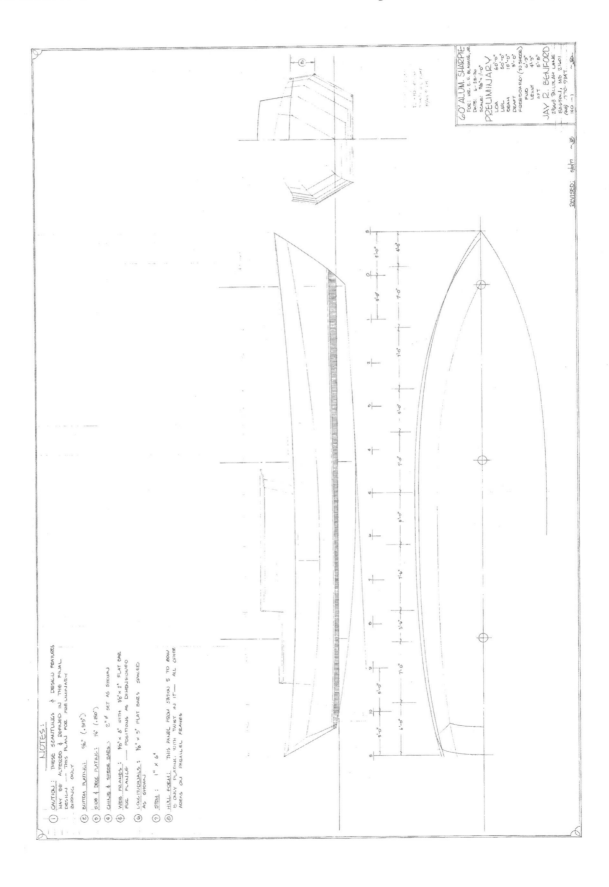

34' Sailing Dory—*Badger*
Design Number 170
1978

The concept of this design was to use the basic interior accommodation plan that worked so well on the 34' Topsail Ketch ***Sunrise,*** which was evolved and improved on during the decade that I lived aboard her. We'd done several sailing dory designs before this one and it benefited from improvements we would make for a better hull form, both for sailing characteristics and for stability.

The oft asked question is not only why is ***Badger*** such a good yacht, but specifically, why is she a good yacht for voyaging on a small income? One of ***Badger***'s greatest attractions is that she is actually designed for just one couple. Most boats of her size have at least six berths and therefore the rest of the accommodation has to be built in around them. On many boats that are used for voyaging, the quarter berths are used for storage. However, on ***Badger***, one can readily gain access to the space under the cockpit, so this area can be used for much more efficient stowage.

Badger is designed to have a large and usable galley, a necessity on any serious voyaging yacht. She has a pleasant saloon with room for bookshelves and a double cabin. The head is large enough so that you can close the door and have a shower. She has a full-width/raised-deck cabin from cockpit to forward cabin, which gives a great sense of spaciousness and is much stronger, structurally, than a conventional coach-roof. There is room for a heating stove. The dory hull gives a wide flat floor, which allows for the accommodation to be pushed further to the sides of the boat without you having to stand on the sides of the hull. She is comfortable to live on both while at sea and in harbor.

Plywood is a simple, quick and strong material with which to build. By shopping around carefully it can be bought for a very reasonable outlay and if you are building while working, it is possible to buy a little at a time. If you are using epoxy, it is not necessary to choose the best quality marine ply—even though that would be first choice—well made exterior can be satisfactory. When epoxy is used, you don't need expensive fastenings, which tend to make up for the initial cost of the glue. Glued and epoxy sealed construction has the advantage that it doesn't leak, a great advantage for any boat. The yacht is of moderate displacement, meaning that the initial building costs are also moderate. An advantage of plywood that is rarely mentioned is that it is very easy to repair, because the damaged area can be cut out and a new piece or pieces scarfed in.

There are two versions of how to build the deck and cabin; one with a trunk cabin and the other with a raised, flush deck like ***Badger***. The latter makes the most sense to me and I would recom-

Table of Particulars:

34' Sailing Dory	Imperial	Metric
Length Overall	34'-0"	10.36 m
Length Datum Waterline	28'-0"	8.53 m
Beam	11'-0"	3.35 m
Draft—fin keel	4'-6"	1.37 m
Draft—long, shallow keel (opt.)	3'-6"	1.07 m
Freeboard, Forward	5'-0"	1.52 m
Freeboard, Least	2'-9"	0.84 m
Freeboard, Raised Deck	5'-1"	1.55 m
Freeboard, Aft	4'-0"	1.22 m
Displacement, Cruising Trim*	10,400	4,717 kg
Displacement-Length Ratio	211	
Ballast	4,160	1,887 kg
Ballast Ratio	40%	
Sail Area, Square Feet	600	55.74 m²
Sail Area-Displacement Ratio	20.15	
Prismatic Coefficient	.56	
Pounds Per Inch Immersion	906	
Auxiliary Horsepower	18	
Water, Gallons	100	378 liters
Fuel, Gallons	40	151 liters
Headroom	6'-3½"	1.92 m

*__Caution:__ The displacement quoted here is for the boat in coastal cruising trim. That is, with the fuel and water tanks filled, the crew on board, as well as the crews' gear and stores in the lockers. This should not be confused with the "shipping weight" often quoted as "displacement" by some manufacturers. This should be taken into account when comparing figures and ratios between this and other designs. Also, loading down for voyaging and living aboard, as with the Hill's *Badger*, will add considerably to these figures, perhaps as much as 50%. Designs like the 34' *Badger*, which load down gracefully and still sail well, make a good choice for anyone wanting to go voyaging on a small income.

mend it for anyone going voyaging. It adds to the room below and makes the deck layout more open and easier to work upon. It also adds to the stability in a knockdown by adding volume where it does the most good in shifting the center of buoyancy in the right direction.

The original cutter rig was done for building in an area where grown poles are available as spars. We worked out some simple hardware that could be made with modest equipment and this rig has worked out well on some sisterships.

As Pete and Annie have found, it is a layout that has worked out very well for them, as it did for me. For anyone wanting to do the sort of cruising that the Pete and Annie Hill are doing, this boat would be hard to beat. It's got room for a couple to live and cruise in comfort and yet is of a size that is affordable and manageable.

On deck, **Badger** is simple and uncluttered, with a small footwell aft, clear center deck and a sunken foredeck which keeps spray away from the cockpit area and allows a solid dinghy to be carried without impeding the helmsman's view. She has plenty of hatches for ventilation. The deck boxes abaft the back of the cabin allow petrol (gasoline—if you must have some aboard) to be stored safely and provide a home for the tails from the sheets and halyards. They also make very comfortable seats. The rudder is hung outboard, for ease of maintenance.

The junk rig is possibly the best short-handed cruising rig ever devised. It is also very inexpensive to build and to maintain. It allows more room below decks and is uncluttered above decks.

Badger can be built simply and for very little money. Sheathed in cloth and epoxy she is easy to maintain and can be kept shipshape and Bristol fashion at very little expense—an essential prerequisite for a boat that is sailed on a small income.

On the following page is shown the LINES plan which shows that **Badger** has a very easily driven shape. She can be gracefully loaded down for a long voyage and still have a clean exit, not dragging an immersed transom through the water. Note the camber curves for the raised deck or trunk cabin and the full size rudder foil sections. The KEEL PLAN shows how to build the fin keel with a rebar cage, covered with welded mesh, and filled with concrete and scrap metal. As the notes indicate, this is a NACA foil section and will give good performance.

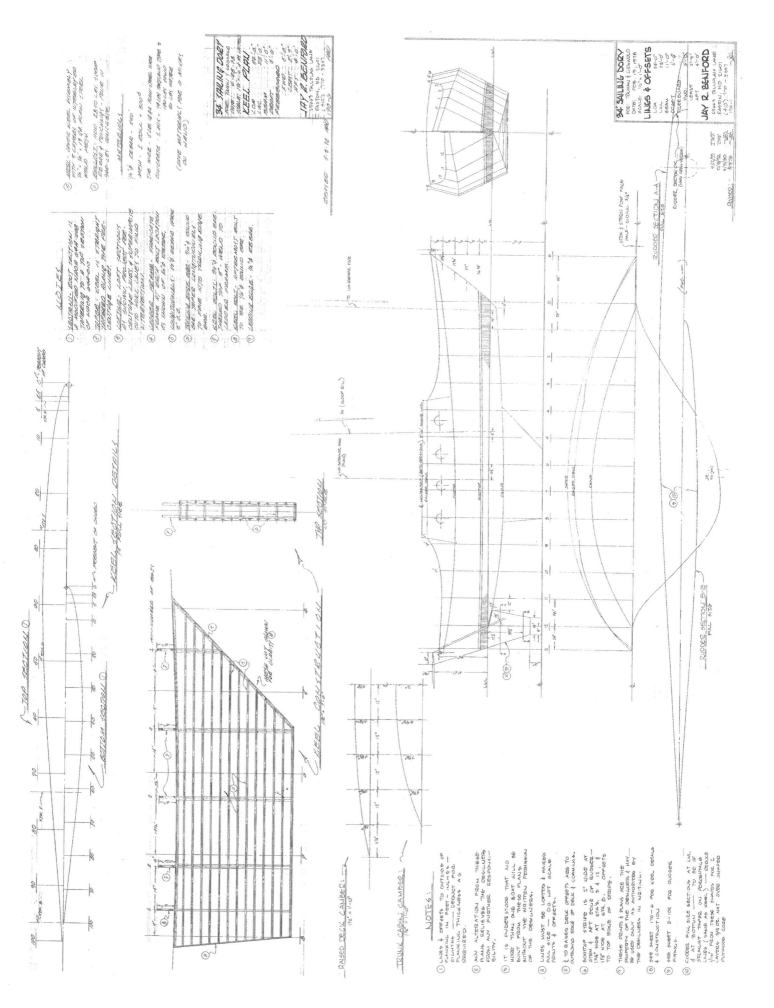

The SECTIONS drawing has larger scale typical midships sections, showing the port side of the trunk cabin version and the starboard side of the raised flush deck version. Also on this sheet are a group of detail sections of how various parts are fitted and scantling notes.

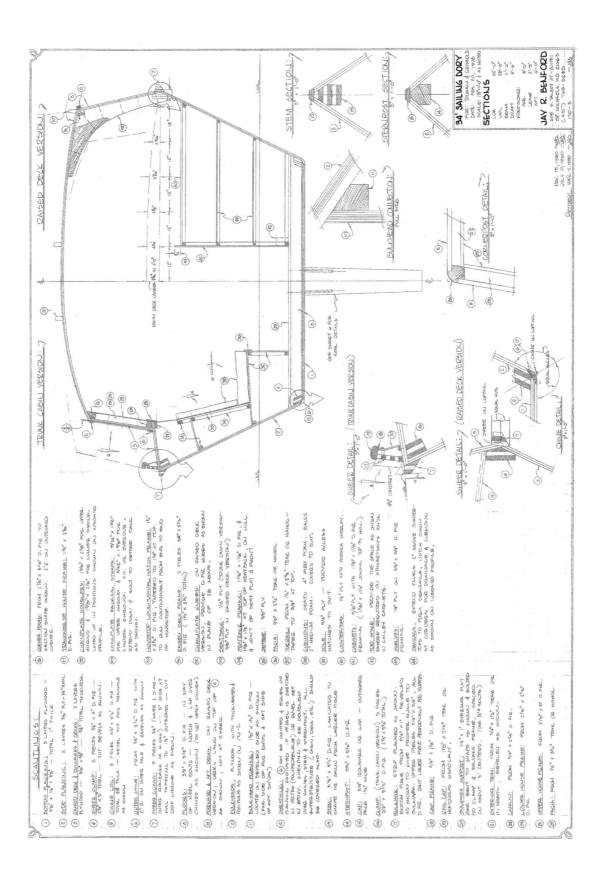

*The original SAIL PLAN is this cutter, with a short bowsprit. The trunk cabin, good side decks and low bulwarks give her a conservative, traditional cruising yacht look. This is very much the layout we had on **Sunrise** that was my home for a decade. Very practical for a couple to live aboard with occasional guests—separate sleeping space, good storage, generous chart table, large head compartment, open saloon and galley space.*

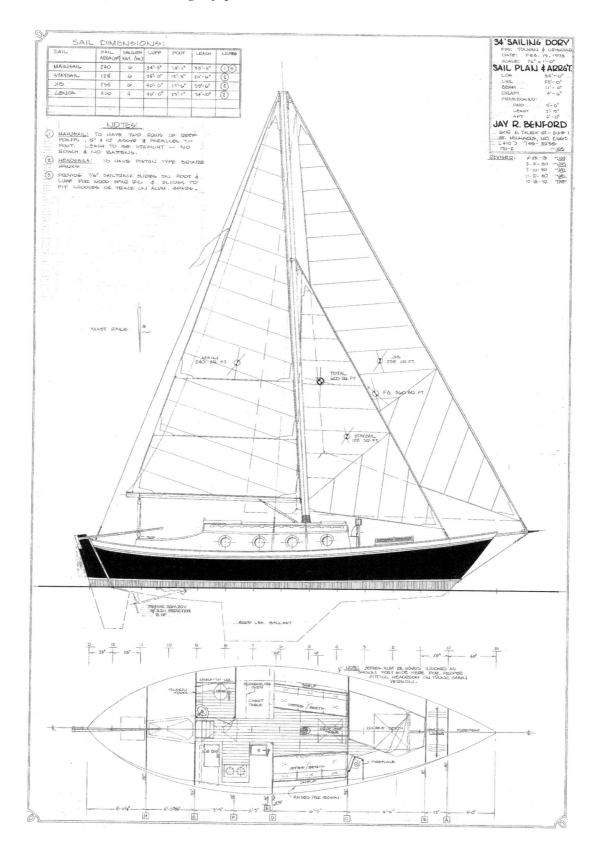

The INBOARD PROFILES sheet has the port inboard profile showing the fin keel cutter and the starboard inboard profile with the long, shoal keel and the junk schooner rig. The lower portion of the drawing has the bottom and deck framing shown on two sides of the plan view. The alphabetical sections are created by using the spacing shown on 170-2 and laying these out when lofting the lines full size.

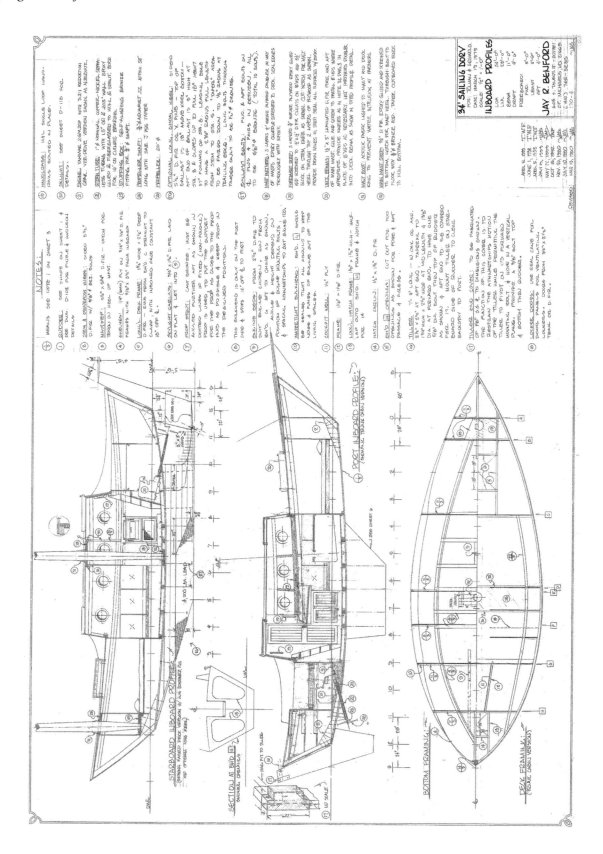

Building *Badger*

The RIG & DECK PLAN has details and notes relating to the cutter rig. The deck plan is split with one side showing the raised deck version and the other side shows the trunk cabin version. Our original client was an Alaskan and this drawing shows solid Sitka Spruce spars. The hardware is intended to be built with some help in machining or welding.

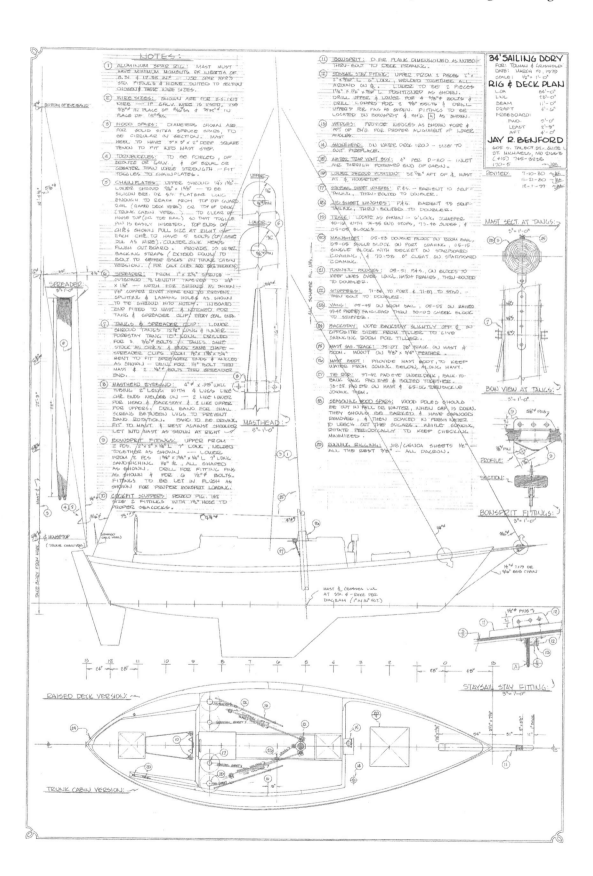

SAIL PLAN & ARRANGEMENT which shows **Badger** as she was built, with the two masted junk schooner rig, the slightly revised accommodation plan and the Collins Keel. This keel is a proprietary design and was purchased by the Hills from a builder who had a couple too many on hand. We cannot, unfortunately, supply the design plans for this particular keel. It's only shown in the interest of accurately portraying **Badger** as she exists.

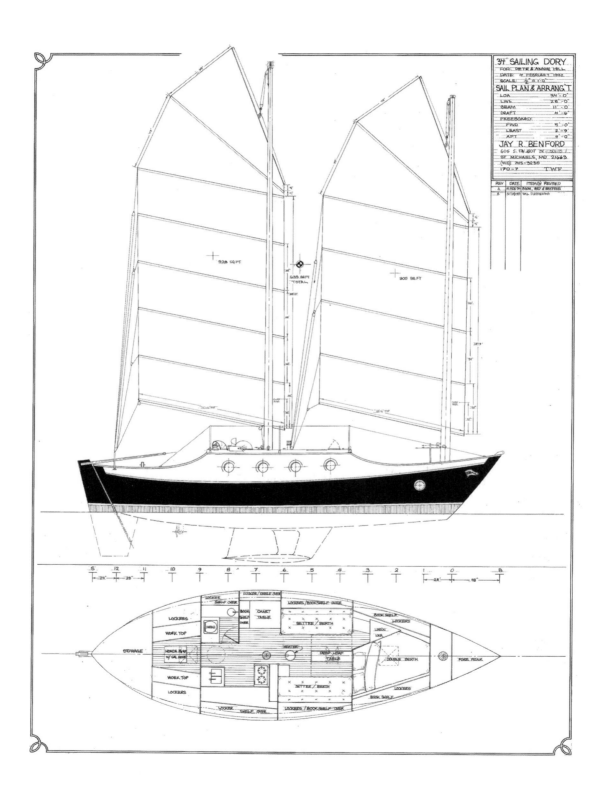

*SAIL PLAN & ARRANGEMENT, the variation requested by a builder who wanted to build **Badger**, but wanted the trunk cabin appearance. This drawing has essentially the same accommodations as the preceding page, with the exception of having a trunk instead of the raised flush deck and the original fin keel instead of the Collins keel.*

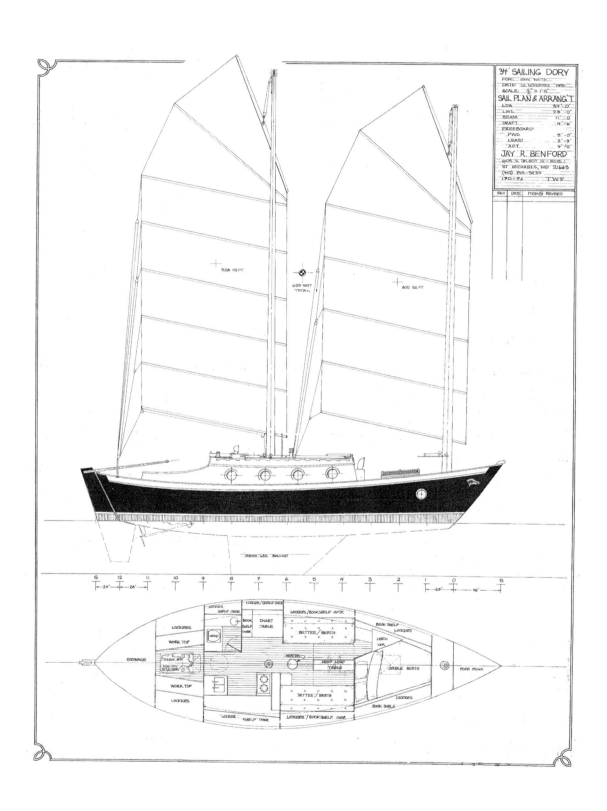

The RIGGING & DECK PLAN for **Badger**. There is a list of equipment and a good layout of the gear on deck so you can see how she was set up for ease of handling.

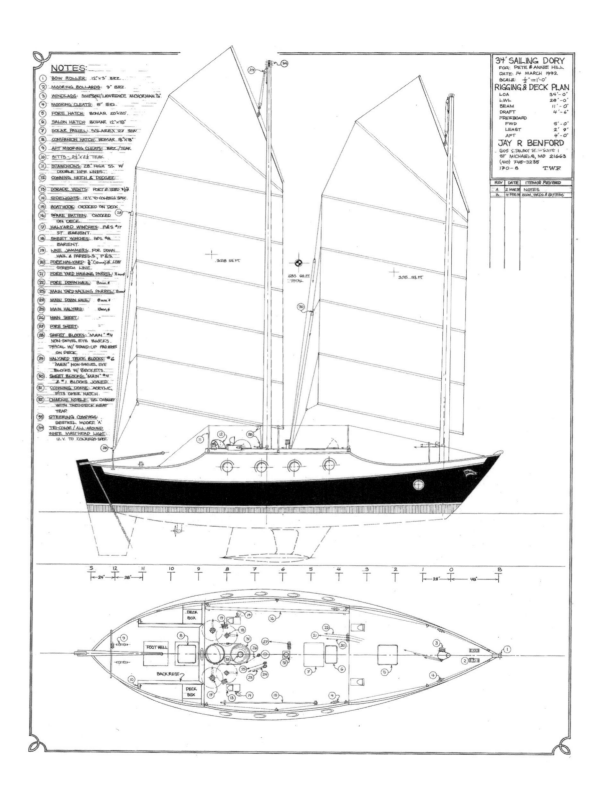

Building *Badger*

*BOOMS, YARDS, & BATTENS—detailing for **Badger**'s—rig is below.*

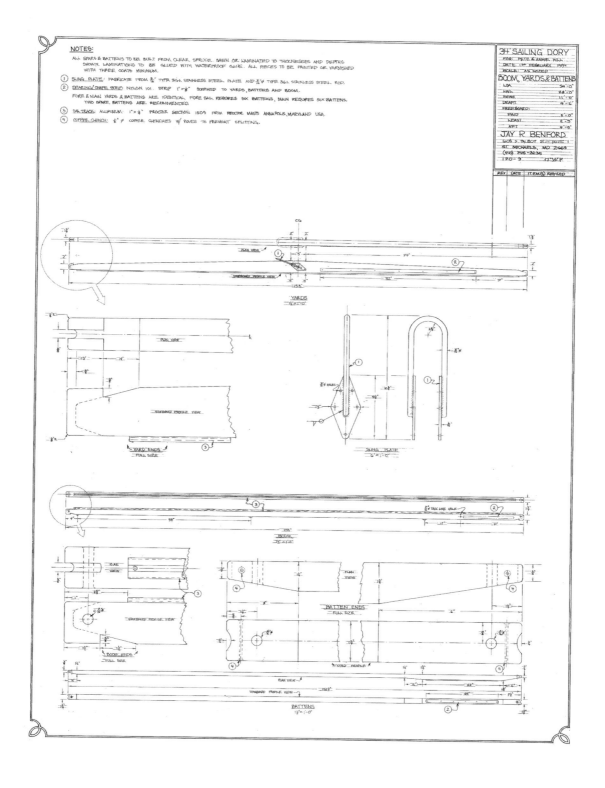

The MAST DETAILS drawing shows how to build hollow spars for **Badger**. **Badger** *was built with solid spars and their weight contributed to part of her floating below her designed sailing lines. We did an inclining test on* **Badger** *to confirm what her stability actually was in her cruising trim. She was sitting 7" deep from her designed waterline at the time, and we found she had a range of positive stability out to 125° of heel even with these heavy spars.*

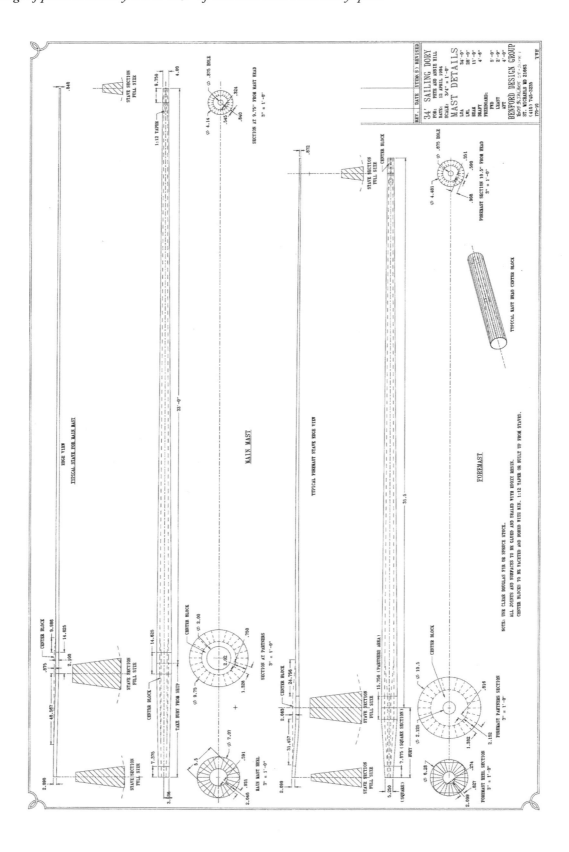

Building *Badger*

The Gaff Cutter SAIL AND DECK PLAN was another request we had for a design variation. Again, we have shown both the trunk cabin and raised flush deck versions on different sides of the deck plan and the section at the mast. And, there is detailing for the bowsprit fittings.

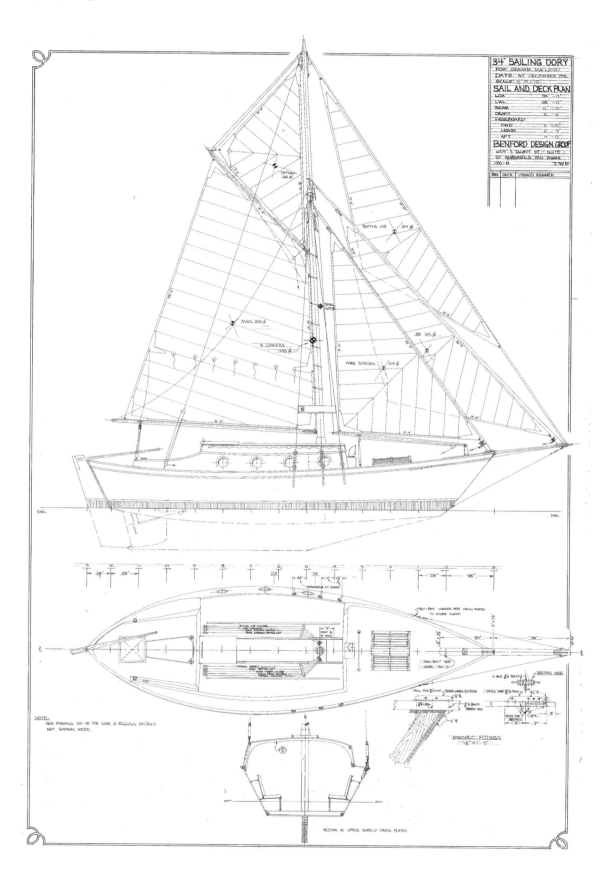

ENGINE & TANKS DETAIL covers a lot of the systems aboard **Badger**. *It addresses the tanks (size, location and construction), piping schematic, engine installation, sight gauges and head installation. The D-sized drawing is a lot easier to read in the actual working plans—four times the size of this page.*

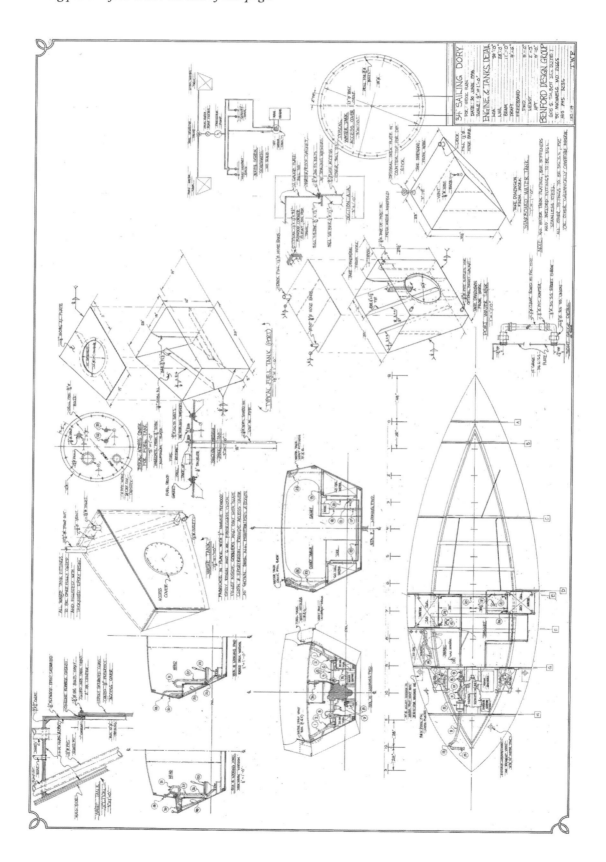

Building *Badger*

*The left photo shows **Badger**'s chart table under construction. It is of generous size and has a lot of storage under it, in the way of drawers and locker space. The right photo shows her forward stateroom, with the double berth and rows of bookshelving and lockers over the berth, built out on the flare of the topsides. Also note the skylight overhead and the natural light coming in. It's great to be able to see what the weather is before even getting out of bed. . . .*

The PILOTHOUSE VERSION (below) shows an unfinished concept for having inside steering on a junk schooner. The access to the aft berths is by sitting on the pilothouse sole, putting your feet into the well by the berth, and swinging around to get onto the berth.

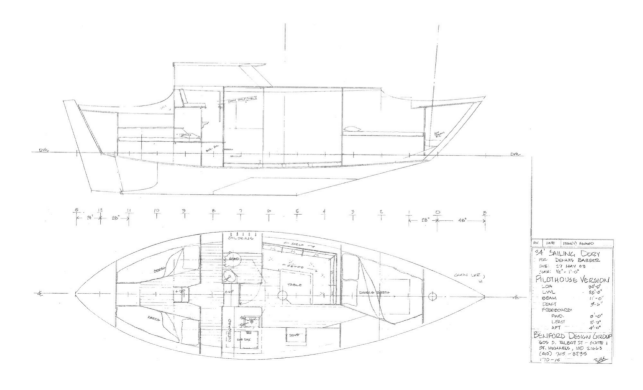

37½' Sailing Dory
Design Number 174
1978

The 37½' design is a modification or variation on the 36-footer, with the stem tilted slightly more forward, a raised flush deck through the middle of the boat, plus a slight length gain at the stern with the extension of the sternpost. The basic keel, bottom and structure are identical, using many of the drawings done for the 36-footer in the construction plan set. The interiors are pretty much interchangeable, and some of the ideas from one can be used in the other size. The trick is to be sure the masts fall in areas that do not complicate the use of the interior space and that the reinforcing for the rigging structure is put in the right place. The raised deck version would again be my recommendation, for the same reasons mentioned about *Badger*, primarily more space and better stability.

There are a couple more variations on the interior layout that are drawn, that are not included here, but are a part of the plans.

Some alternatives to the layouts shown that would be practical would include the addition of a hard dodger or pilothouse to the forward end of the cockpit. If this was done with a version with an aft cockpit and quarter berths, the berths could be shifted aft and the pilothouse covering up this part of the accommodations. Note that the trunk cabin version has the same layout as the 36-footer, with the exception of having a double berth in the bow.

37½' Sailing Dory

The Lug Schooner version of the 37½' dory, has an aft cockpit. The lug rig means simplified sail handling, but is not as easy to add sail area to it in light airs. The alternate interior, below, moves the galley aft and has a quarter berth instead of a workbench. The 37½' Cutter has a midships cockpit, recessed into the raised deck. This leaves room enough to carry an 8 to 9' hard dinghy on the aft deck, over the giant skylight which is over the dining lounge.

Right: Anthony Swanston's 37½' **Wild Fox** *of Belfast, Northern Ireland, ready to go overboard for the first time.*

Building *Badger*

Table of Particulars:

37½' Sailing Dory	Imperial	Metric
Length Overall	37'-6"	11.43 m
Length Datum Waterline	31'-0"	9.45 m
Beam	11'-0"	3.35 m
Draft—fin keel	4'-6"	1.37 m
Draft—long, shallow keel (opt.)	3'-6"	1.07 m
Freeboard, Forward	5'-2½"	1.59 m
Freeboard, Least	3'-0¼"	0.92 m
Freeboard, Raised Deck	5'-4"	1.63 m
Freeboard, Aft	4'-9"	1.45 m
Displacement, Cruising Trim*	13,425	6,008 kg
Displacement-Length Ratio	201	
Ballast	5,350	2,247 kg
Ballast Ratio	40%	
Sail Area, Square Feet	700/725	65/67 m²
Sail Area-Displacement Ratio	19.83/20.54	
Prismatic Coefficient	.63	
Pounds Per Inch Immersion	983	
Auxiliary Horsepower	27	
Water, Gallons	110	416 litres
Fuel, Gallons	55	208 litres
Headroom	6'-4"	1.93 m

Caution: The displacement quoted here is for the boat in coastal cruising trim. That is, with the fuel and water tanks filled, the crew on board, as well as the crews' gear and stores in the lockers. This should not be confused with the "shipping weight" often quoted as "displacement" by some manufacturers. This should be taken into account when comparing figures and ratios between this and other designs. Also, loading down for voyaging and living aboard, as with the Hill's ***Badger***, will add considerably to these figures, perhaps as much as 50%. Designs like the 34' ***Badger***, which load down gracefully and still sail well, make a good choice for anyone wanting to go voyaging on a small income.

*More launching photos of Anthony Swanston's 37½" **Wild Fox**. The photo at right gives a good view of her underwater shape and the photo above shows her raised deck topsides.*

LINES & OFFSETS drawing can be compared to the 36-footer and see how closely they are related.

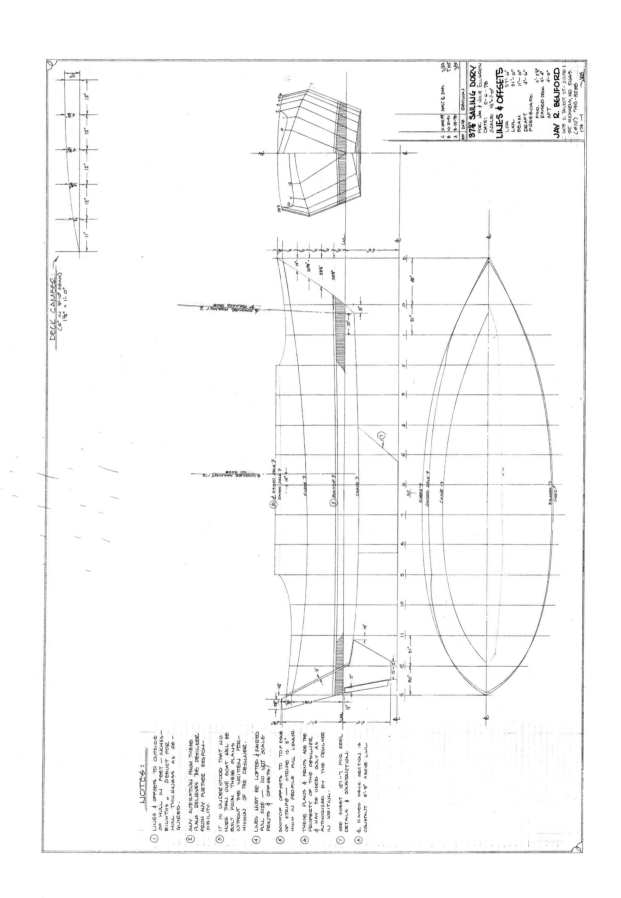

Building *Badger*

The SAIL PLAN & ARRANGEMENT drawing shows the lug (junk) schooner rig. A half century ago, this style was called a lug rig—standing lug in this case, as opposed to a dipping lug in which the lug is dipped around the mast when tacking. The popularization of the junk rig with the long-distance single handed racers has now made the Eastern name for the rig more common than the Western one. The arrangement shown has the galley forward of the saloon, with the engine in a box under the dining table.

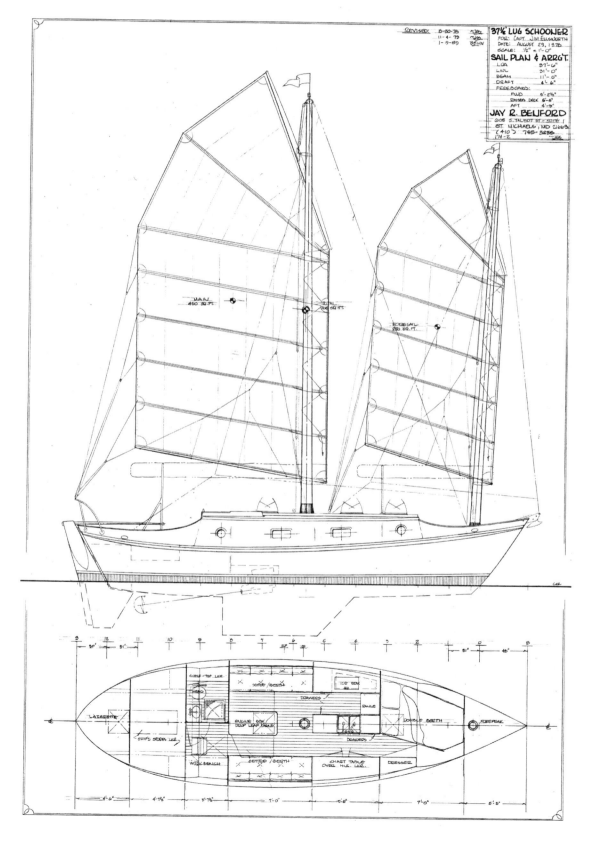

The CONSTRUCTION is laminated plywood over longitudinal frames at the chine, sheer, and raised deck with transverse bulkhead framing and floors to help spread the load of the ballast keel. On the large section, you can see the longitudinal deck framing above the deck, which provides framing for the hatches and mast reinforcement areas.

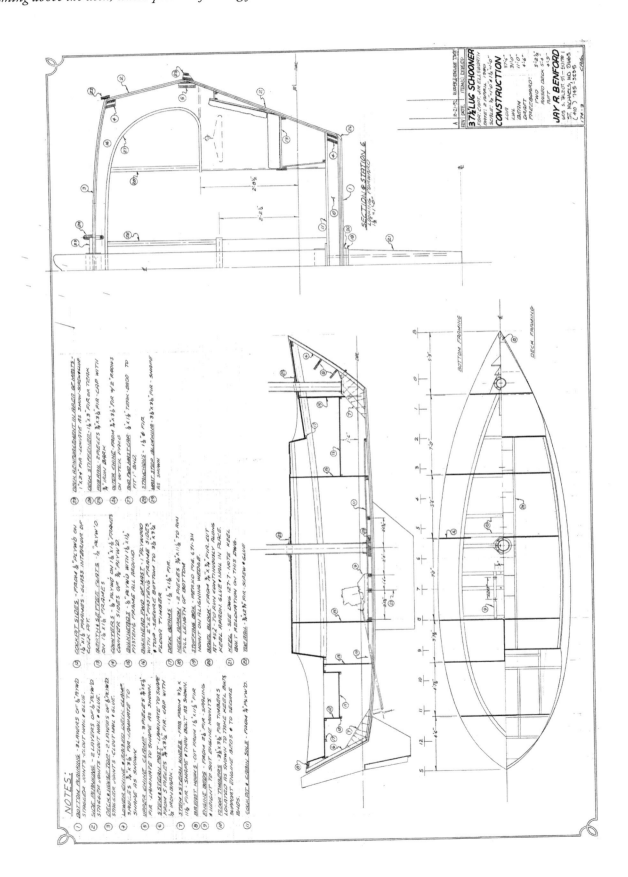

The Cutter version's SAIL PLAN & ARRANGEMENT shows the midships cockpit, great cabin aft layout. The big skylights over the forward berth and the aft saloon will let in lots of light and give the option of having fresh air when weather conditions permit. The passageway under the cockpit has less than full headroom where it passes under the cockpit seat, but does provide fore and aft passage without going on deck. The panel of the engineroom along this passageway would be an opening one to permit access and servicing of the engine.

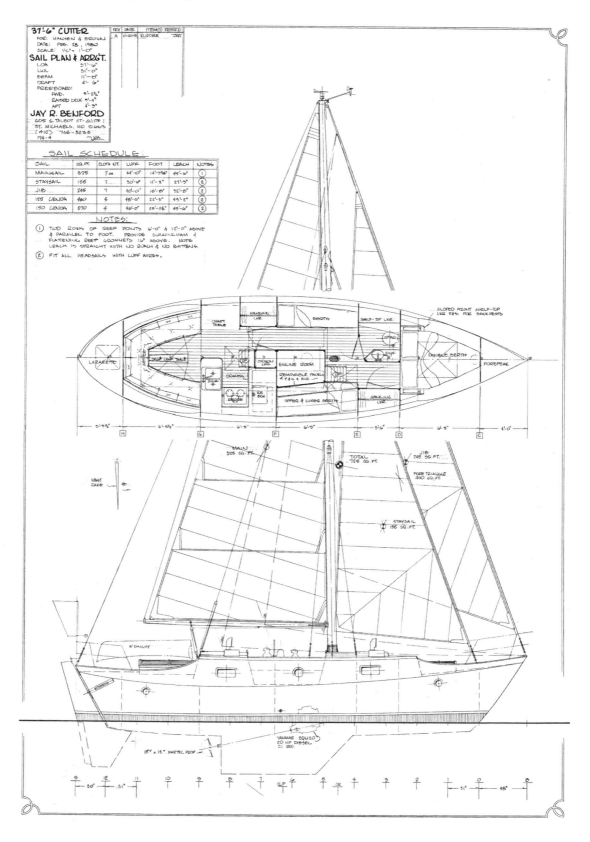

The SAIL PLAN & ARRANGEMENT #2 is also a cutter, but this time with a trunk cabin and layout almost like the 36-footer, with the exception of having a double berth in the bow. This layout has the features of the 34, with the addition of the quarter berths. This makes for a good way to have children living aboard, or for taking extra crew on a passage.

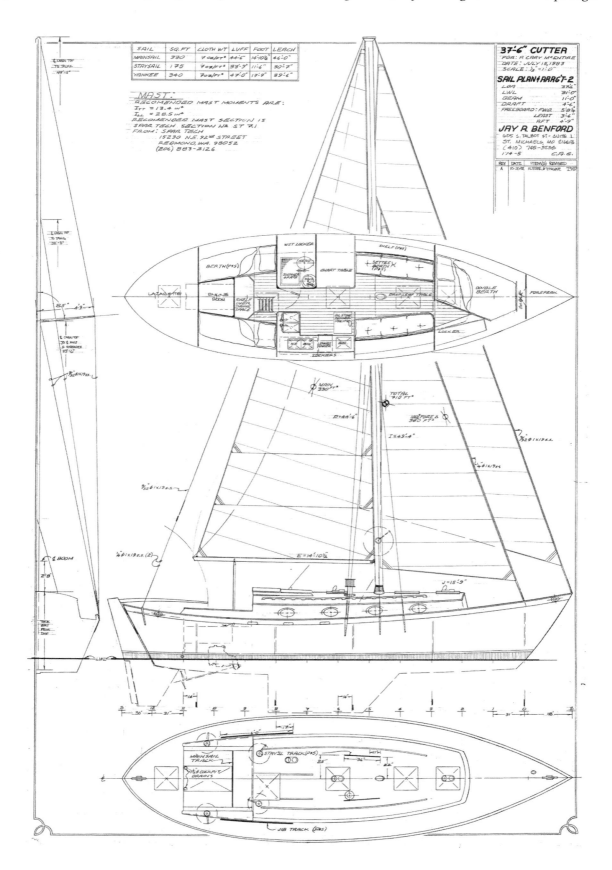

The RIG & DECK PLAN for the junk schooner lays out the run of the lines as the cross the deck and has dimensions for the sails.

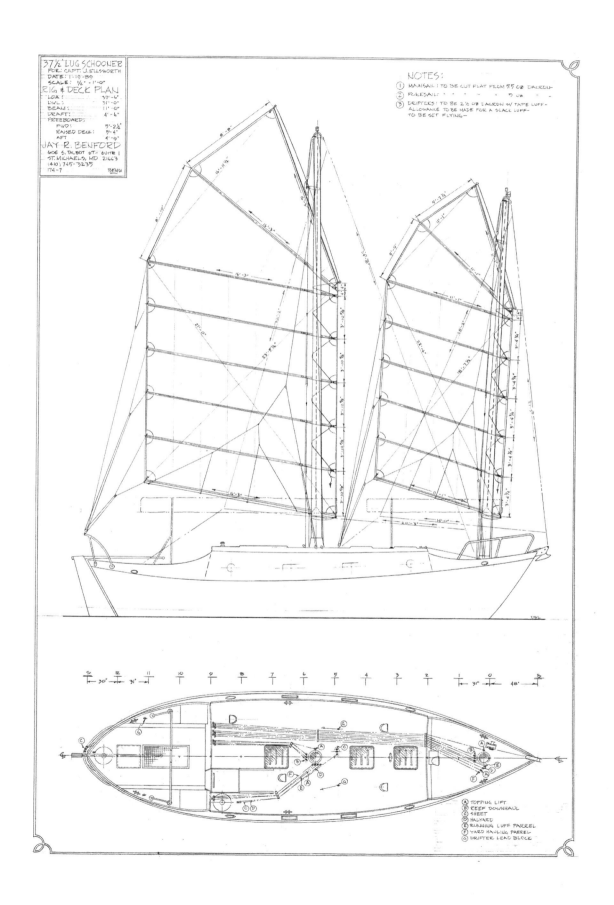

LONG KEEL VERSION shows the 3'-6" draft keel. This variation uses lead instead of concrete and scrap metal to keep the center of gravity low and the stability high.

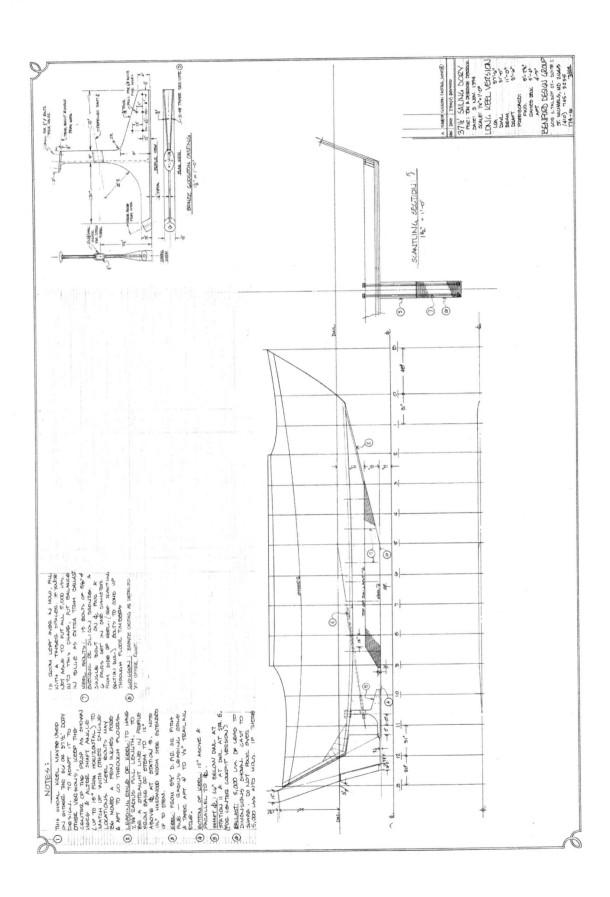

ACCOMMODATIONS *shows how to make a center cockpit layout with the junk rig. There are two entries to the cabins, a companionway hatch and steps going aft from the cockpit and just forward of the cockpit an on deck hatch and ladder into the forward cabin.*

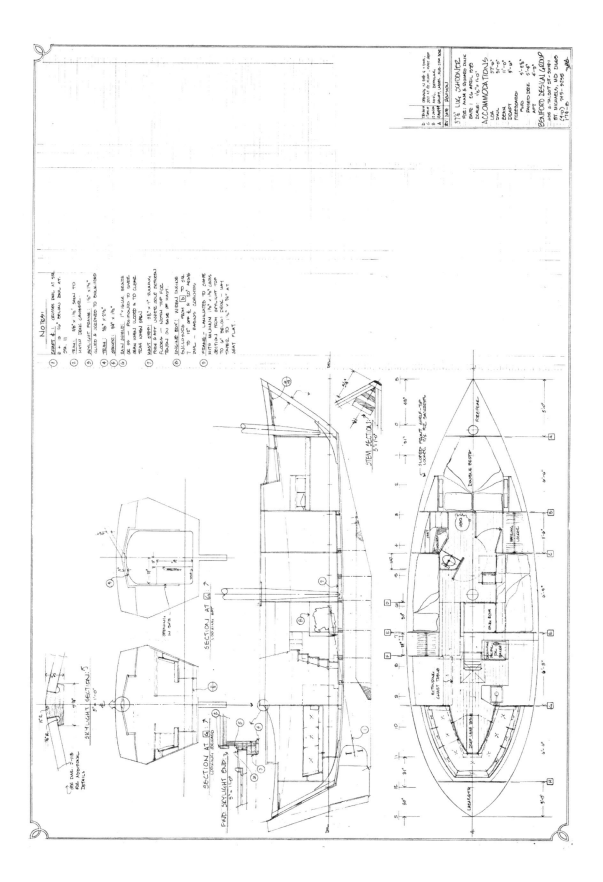

DECK PLAN & DETAILS has the revisions to the plans related to the layout shown on page 155. That belaying pin rack across the front of the cockpit will add a salty touch to her appearance. Note the sloped seatbacks in the cockpit, making that a comfortable place to spend long periods of time, whether on watch or at anchor.

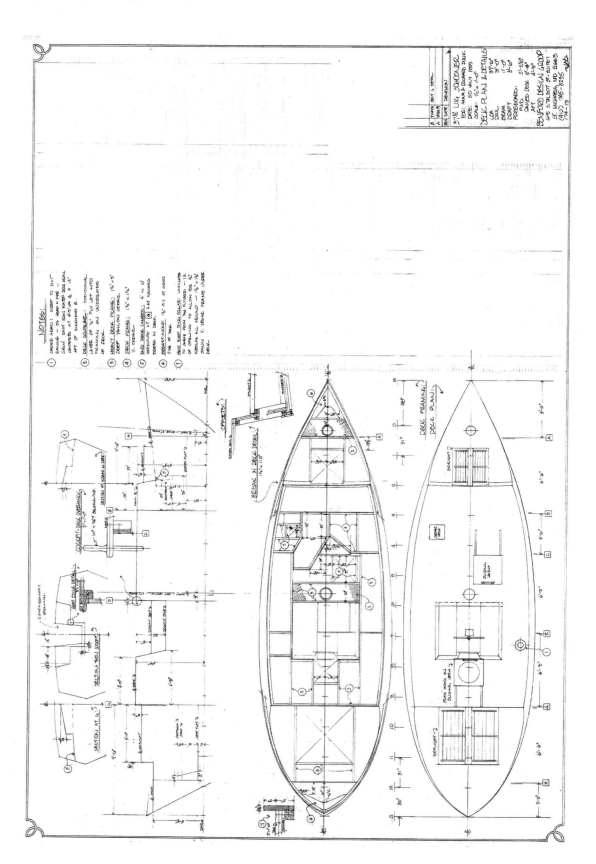

The Dory Trawler version's PROFILE & ARRANGEMENT shows an altogether different version of this design. She would have 3,000 miles range under power alone and the sails would give her a lift and cut fuel consumption on long passages—and steady her motion in a seaway. The layout is similar to the midships cockpit variation shown on 174-18, except for the pilothouse where the earlier one has passage berths in this area. The pilothouse has steps down to the forward head and stateroom, and down to the aft galley, chart table and dining saloon. As the section shows, there are passage berths tucked under the pilothouse seats and the side decks.

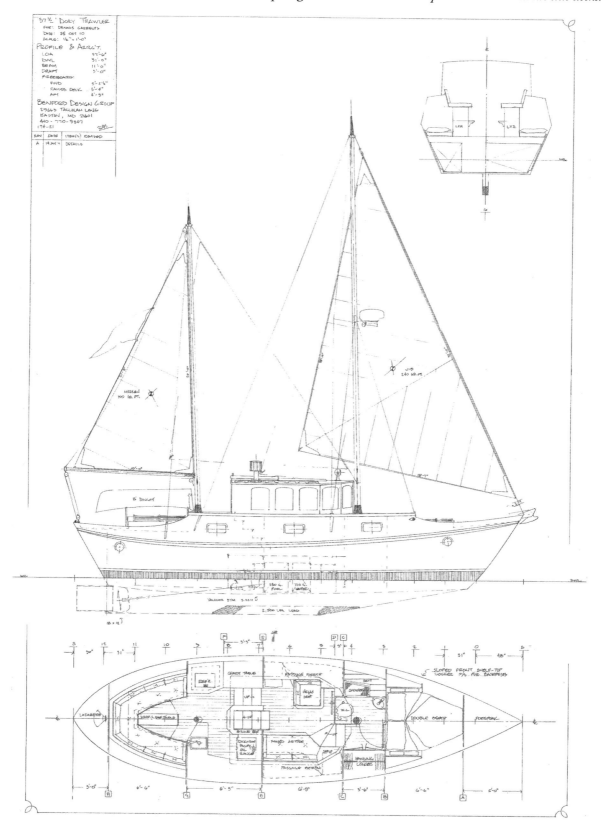

23' Knockabout *Sourdough*
Design number 179
1979-81

This design was commissioned by an Alaskan — hence her name. He, being quite adventurous, planned to use her for long distance cruising, perhaps making runs to places like Hawaii. Therefore, she had to be suited to ease of operation for singlehanding and have good range under power.

The hull form is an evolutionary development from quite a number of our previous dory designs. We've had very good reports on the seakeeping abilities of these flat-bottomed craft. When sailing, they're heeled and presenting a v-shape to the water so they do not pound. At anchor, the chine forward is immersed and will not slap or pound in the normal light chop found in a sheltered anchorage.

Construction is plywood planking over fir longitudinals and bulkheads, using epoxy gluing and sealing. Every reasonable step has been taken to make this as simple and straightforward to build as possible, and still maintain a vessel with good seaworthiness. There will be very few pieces required for the construction that will not be available from a good lumberyard or hardware store.

There is virtually no twist required in putting on the plywood planking. The forward deck has its centerline straight and parallel to the designed waterline. Thus, with only camber and not sheer to follow, the decking will lay on easily. The camber on the deck gives the sheer seen in profile. The longitudinal frame on the companionway slide and forward hatch is one long piece that goes forward and laps over the forepeak bulkhead. This nicely stiffens the deck, leaving the underside clear for maximizing headroom.

The ballast in concrete and scrap metal cast to shape and bolted in place. The tiller steering with the outboard rudder can be supplemented with the trim tab shown. hooked to a wind vane steerer.

The knockabout rig was our first thought for a very simple and easily assembled rig. It has only three wires holding up the mast, and the spars are grown sticks. Later on, our client thought he might like to try a lug rig on her, and we laid out the lug rig shown on another drawing. Both rigs can be handled from aft, and the only reason to go forward will be for anchor handling.

The cockpit is built with good height and slope on the backrests to provide a place where the crew can spend long periods of time in comfort. The cockpit sole has a slight slope aft, and drains through two slots in the transom. This saves several pieces of expensive plumbing and a simple rubber flap over the slot will keep out most of the potential back-flooding. The slots being almost a foot above the water, it is not likely much water will try to come back through them.

The engine originally specified is a Sabb 8hp diesel, with a feathering variable pitch propeller. The ability to feather the prop will minimize drag under sail, and the variation of pitch available will make for efficient

Particulars:	Imperial	Metric	Wide Vers.	Metric
Length overall	23'-0"	7.01 m	23'-0"	7.01 m
Length designed waterline	19'-6"	5.94 m	19'-6"	5.94 m
Beam	8'-0"	2.44 m	9'-0"	2.74 m
Draft	3'-0"	0.91 m	3'-0"	0.91 m
Freeboard: Forward	4'-0"	1.22 m	4'-0"	1.22 m
Least	2'-9"	0.84 m	2'-9"	0.84 m
Aft	3'-3"	0.99 m	3'-3"	0.99 m
Displacement, cruising trim*	4,900 lbs.	2,223 kg.	5,500 lbs.	2,223 kg.
Displacement-length ratio	295		331	
Ballast	1,000 lbs.	454 kg.	1,000 lbs.	454 kg.
Ballast ratio	20%		20%	
Sail area—knockabout	236 sq. ft.	21.93 m²	236 sq. ft.	21.93 m²
Sail area-displacement ratio	13.09		12.12	
Sail area—lug rig	272.5 sq. ft.	25.32 m²	272.5 sq. ft.	25.32 m²
Sail area-displacement ratio	15.11		13.99	
Sail area—Motorsailer	224 sq. ft.	20.81 m²	224 sq. ft.	20.81 m²
Sail area-displacement ratio	12.42		11.50	
Prismatic coefficient	.600		.600	
Water tankage	Gals.		Gals.	
Fuel tankage	20 Gals.	76 liters	20 Gals.	76 liters
Headroom	4'-9"	1.45 m	4'-9"	1.45 m

*CAUTION: The figure for displacement quoted here is for the boat in cruising trim. That is, with the fuel and water tanks filled, the crew on board, as well as the crews' gear and stores in the lockers. This should not be confused with the "shipping weight" often quoted as "displacement" by some manufacturers. This should be taken into account when comparing figure and ratios between this and other designs.

long-range motorsailing. (There are a good variety of other small diesels that would be suitable power for this boat—7 to 10 horsepower is all that this boat will need.) The hatch over the engine provides quick access for hand-cranking the engine, as well as for checking on its health. The side cockpit hatches give access to the storage under the cockpit as well as a method of side access to the engine.

The accommodation plan is one we've found successful on boats this size. The large double berth forward can be partitioned off with a leecloth at sea. The lounging seats port and starboard are shaped and sloped to be comfortable places to spend long periods of time. Cooking and dishwashing can be done from the seats. The stove specified is a wood-burning model, with an over that will take a 9" by 9" baking pan. The warmth and dry heat of this stove will surely endear it to the crew in Alaska. It can be supplemented by a gimbaled one burner while at sea or in hotter weather.

With the specified 20 gallons capacity of fuel, she should have a range of over 600 miles at 5 knots, and about 300 miles at 6 knots. It's possible to increase the tankage, for extended cruising.

Pilothouse Version—Before building the boat, our client took our suggestions about building an enclosed steering station, the results of which are shown on the drawings. As with the original, the rig is modest in size, as fits a boat that will power through the light airs, and not have to reef as soon offshore. Other than the actual changing of the smaller for the larger headsail, all the sailing evolutions are handled from the cockpit. The one sheet winch will handle both port and starboard headsail sheets from its central position.

The interior layout is meant for mostly singlehanding and passagemaking. The berths are moved a bit aft to an area of a little easier motion, and are provided with backrests to make them comfortable places to sit and read as well as sleep. The main item of equipment for the galley is a diesel range with an oven. Besides providing hot meals, this will also heat the boat and make year round cruising more pleasant. For tropical use, a gimbaled primus can be fitted.

With 6'-6" headroom, the pilothouse will be roomy. Windows all around make it easy to keep track of the world going by, and the overhead hatch will make it easier to watch the sail trim.

The pilothouse version has lead ballast to lower the center of gravity of the ballast, offsetting the increased weight aloft of the house. Building the 9' beam version would also make her stiffer— 42% more stability — since that varies with waterline beam cubed. Her cockpit is open, with no seats, so fishing will be easier with the rails more accessible, and there is room for a deck chair or two.

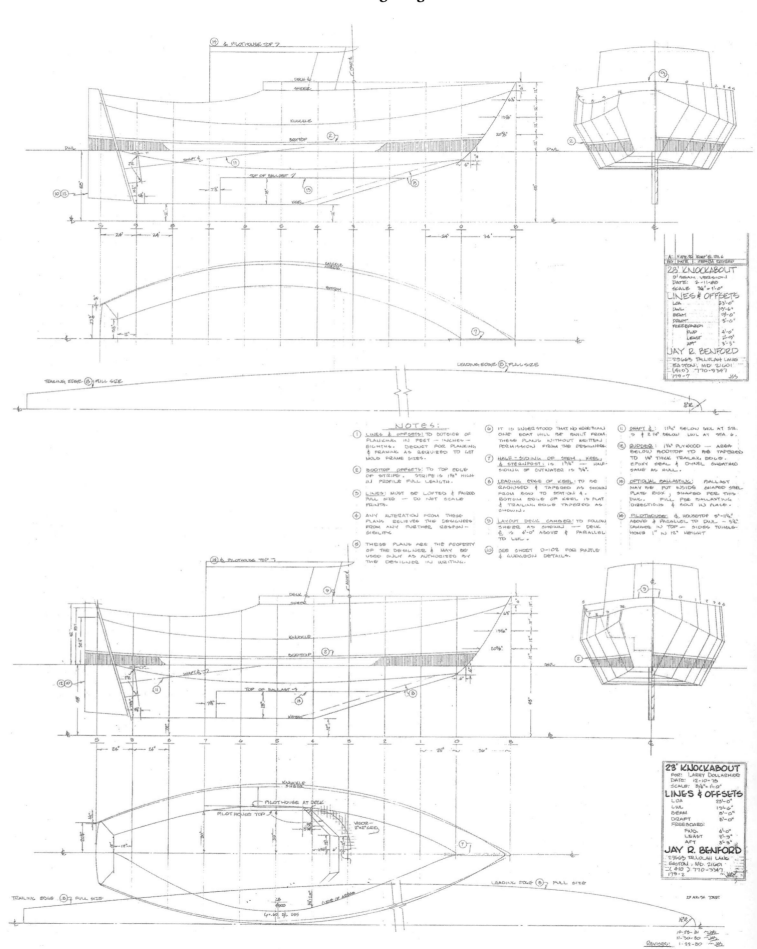

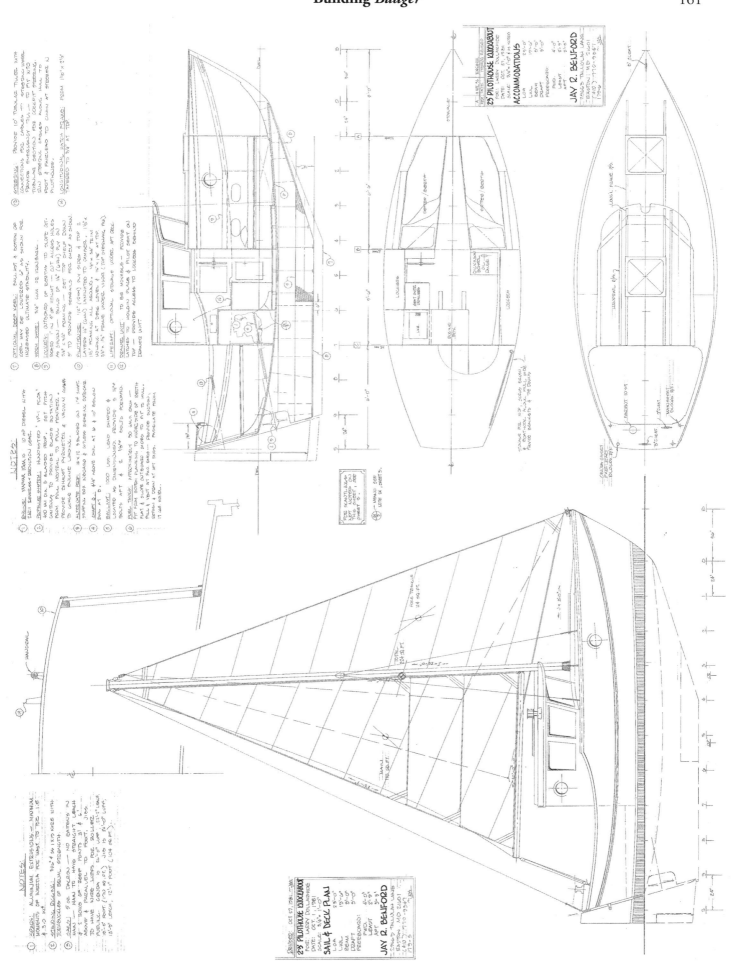

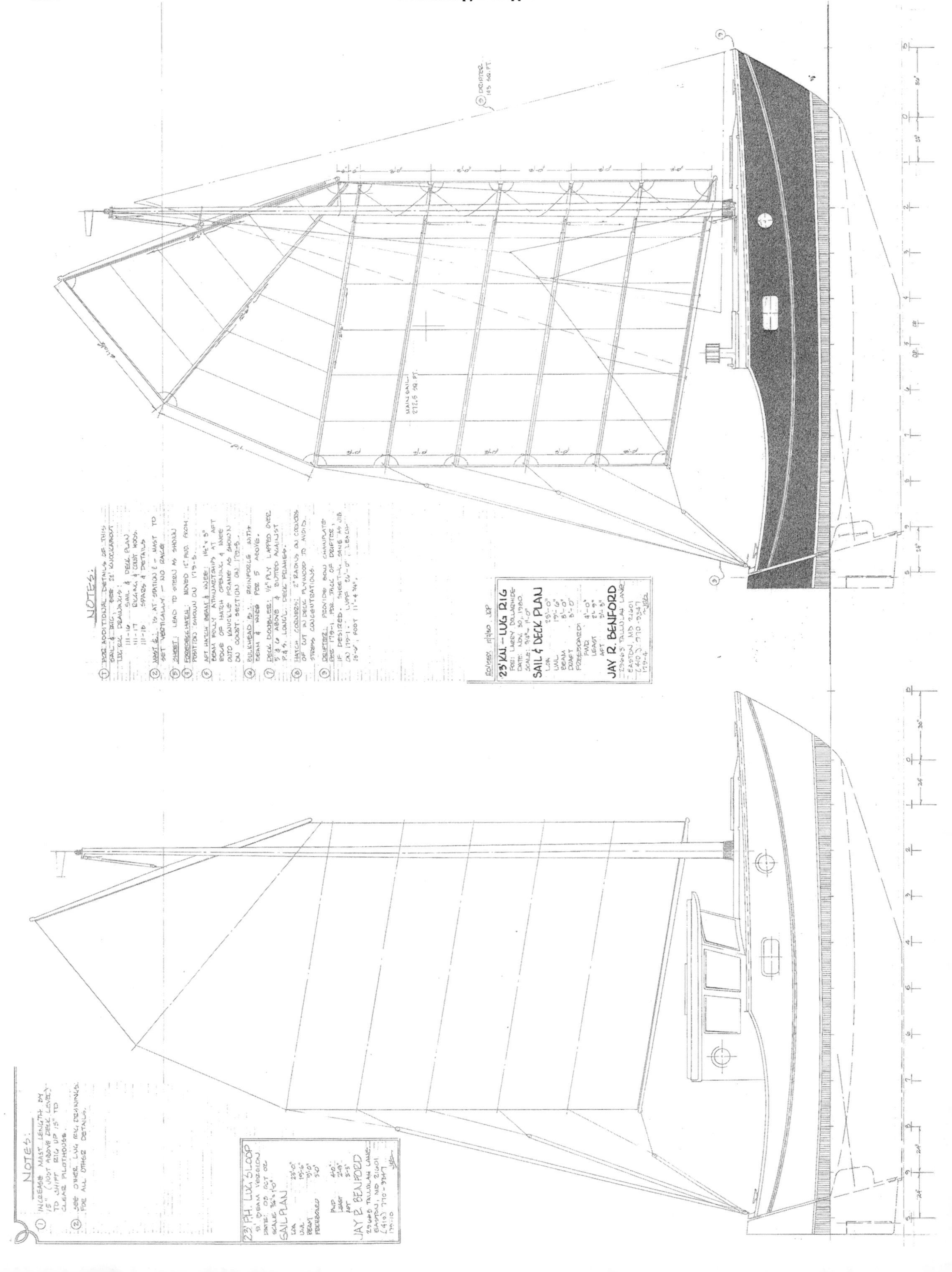

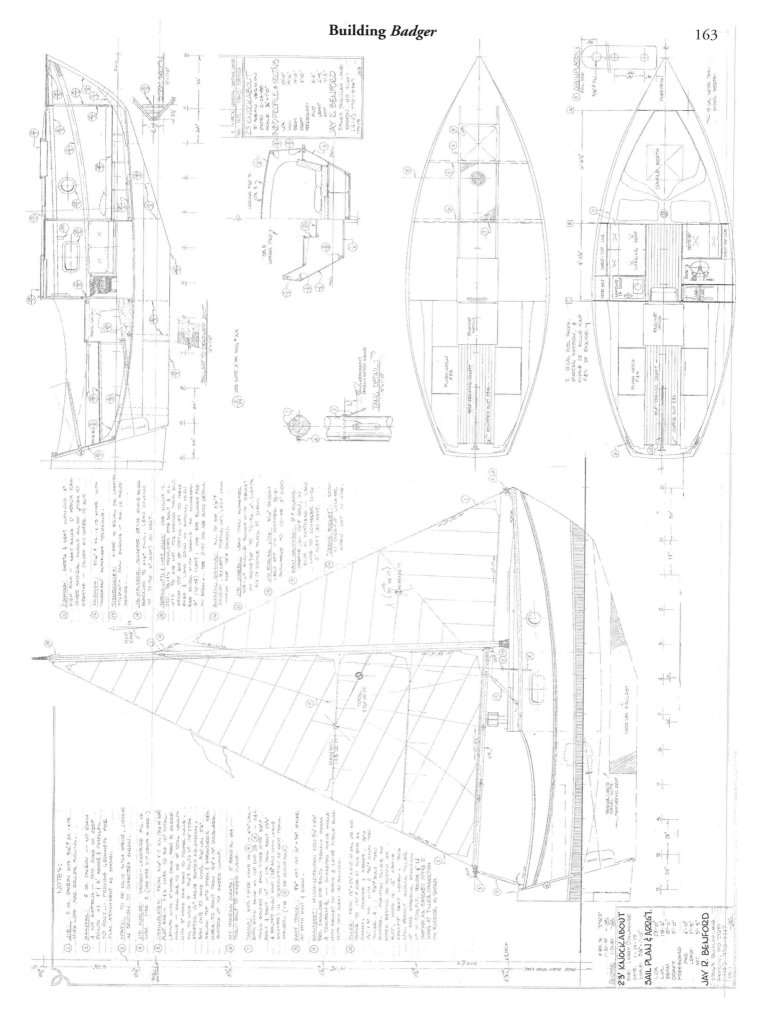

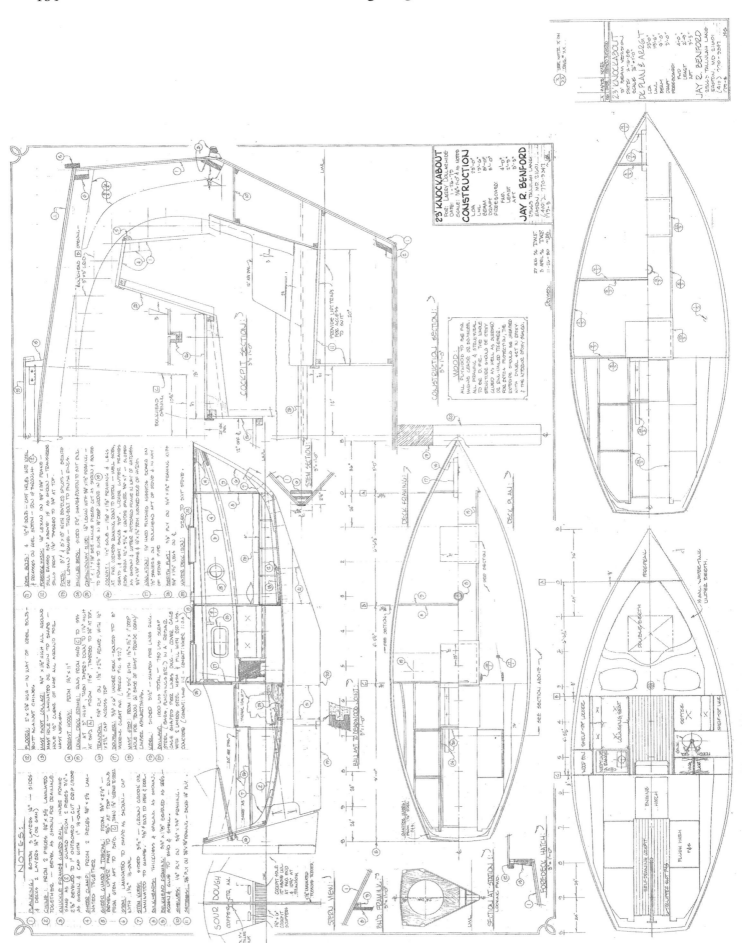

26' Raised Deck Cutter
Design Number 274
1988

Created several years after we did the 23' Raised Deck Knockabout, **Sourdough**, this bigger version has more room greater seagoing ability. The twin keels and centerline skeg and rudder provide three-point grounding ability, for drying out sitting upright, making exploring the shallows more fun.

Construction is plywood over fir framing, glued and sealed with epoxies, for simple and rugged construction, which can be quickly accomplished. She has a small diesel for those times when there's no wind or you want to motorsail.

This design is capable of doing the sort of voyaging that the 34' Sailing Dory **Badger** does. She has comfortable accommodations for a couple to cruise aboard. What she won't have is the sort of extended stores and long-term supplies carrying ability that makes the larger **Badger** work so well. She was designed with full foam floatation, intending to be the sort of boat that would float even if filled with water. (The drawing showing these details is on the top of page 170.) The concern with this for a voyager is that the floatation material is filling a lot of valuable locker spaces, and, if I were doing one for myself, I would bypass the foam floatation in favor of having more stowage space. She will take the ground gracefully when drying out on a tidal cycle, sitting upright on her twin keels and skeg on level bottoms. This can make exploring the shallows and backwaters more relaxing, knowing you won't have to spend a tidal cycle lying well over.

Table of Particulars:

45' Sailing Dory	Imperial	Metric
Length Overall	45'-0"	13.72 m
Length Datum Waterline	39'-1"	11.91 m
Beam	16'-0"	4.88 m
Draft—long, shallow keel (opt.)	4'-0"	1.22 m
Freeboard, Forward	6'-6"	1.98 m
Freeboard, Least	4'-3"	1.30 m
Freeboard, Aft	5'-3"	1.60 m
Displacement, Cruising Trim*	31,800 lbs	14,424 kg
Displacement-Length Ratio	238	
Ballast	10,000 lbs	4,536 kg
Ballast Ratio	31%	
Sail Area, Square Feet	1,209	112.32 m²
Sail Area-Displacement Ratio	19.27	
Prismatic Coefficient	.617	
Pounds Per Inch Immersion	1,603	
Auxiliary Horsepower	60	
Water, Gallons	300	litres
Fuel, Gallons	150	litres
Headroom	6'-6"	1.98 m

Caution: The displacement quoted here is for the boat in coastal cruising trim. That is, with the fuel and water tanks filled, the crew on board, as well as the crews' gear and stores in the lockers. This should not be confused with the "shipping weight" often quoted as "displacement" by some manufacturers. This should be taken into account when comparing figures and ratios between this and other designs. Also, loading down for voyaging and living aboard, as with the Hill's **Badger**, will add considerably to these figures, perhaps as much as 50%. Designs like the 34' **Badger**, which load down gracefully and still sail well, make a good choice for anyone wanting to go voyaging on a small income.

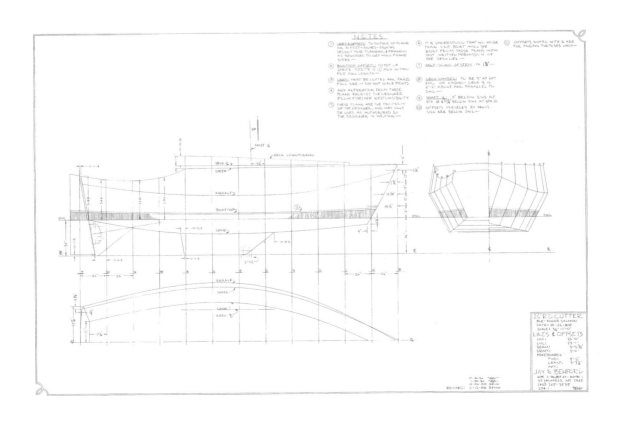

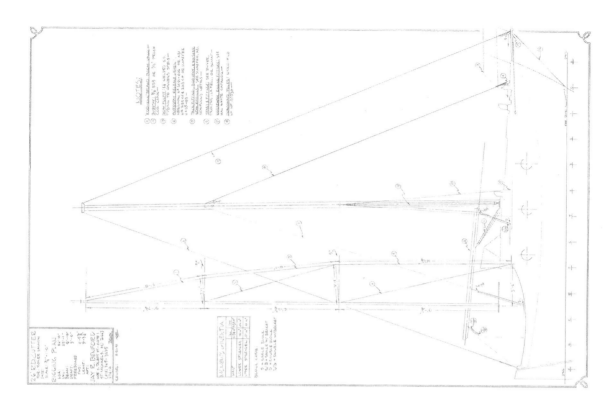
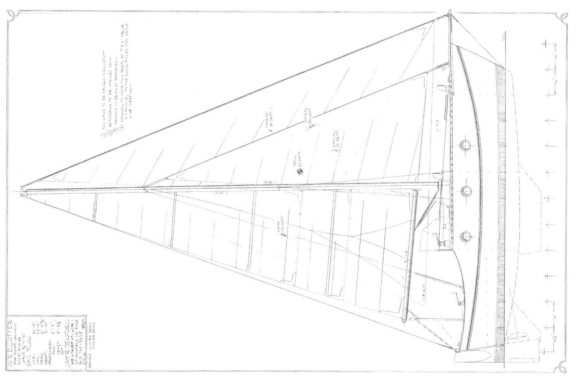

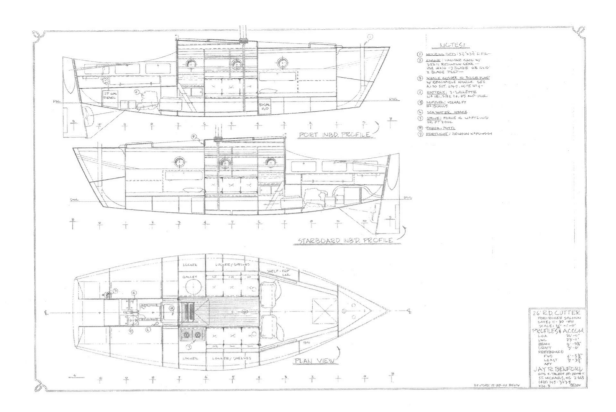
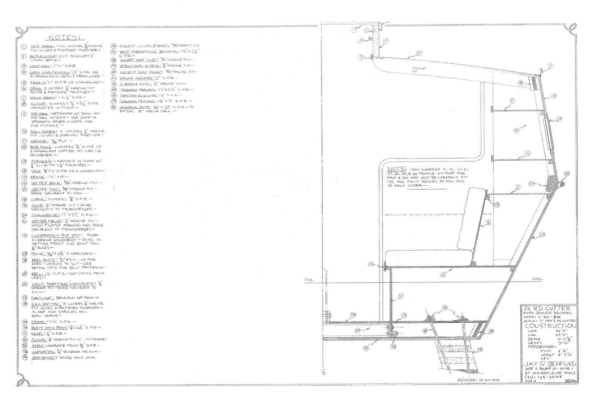

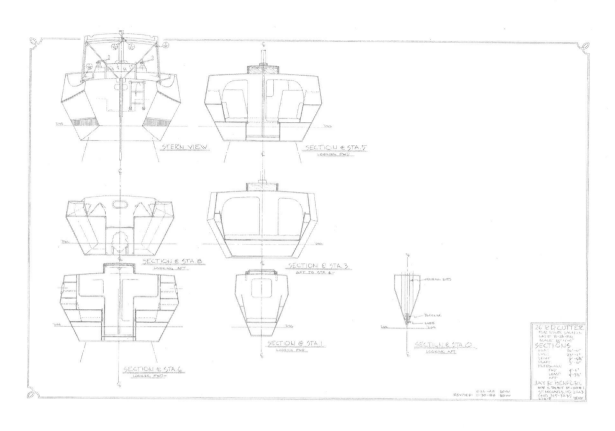

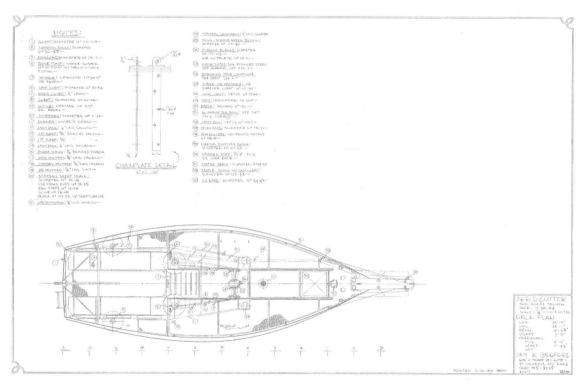

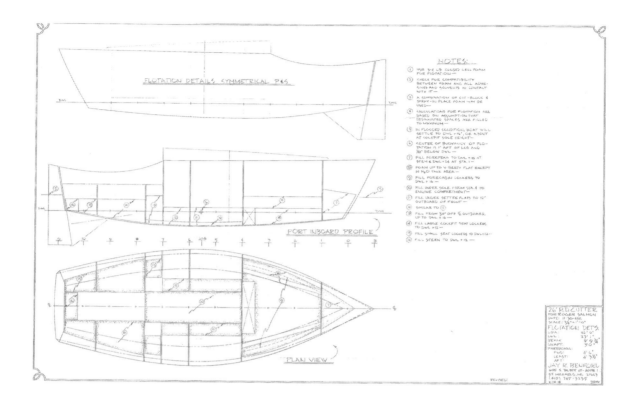
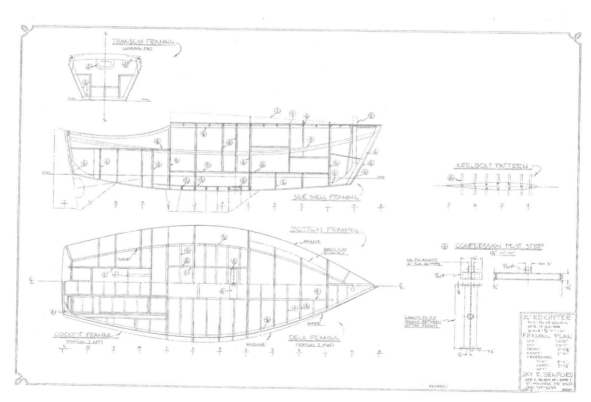

36' Power Dory
Design Number 301
1990

This design has the same hull form as the 36' Sailing Dory. However, instead of the 5,000 pounds of ballasted keel, she has a steel structure and no ballast. In practice, this change in structure worked out to the difference in weight between building the whole structure in plywood and building all in steel. Trying to build the 36 in steel as a sailing boat **and** put ballast on her would result in her being seriously overweight. She could, however, be built in plywood or aluminum like the original sailing version, with a significant reduction in weight and having some ballast added to make up some of the difference.

Two alternative pilothouse styles are shown on the drawings. The windows raked forward at their top will have less reflected internal light at night, making this a practical choice.

She was created for use as a fishing vessel, having an insulated cargo hold under the cockpit. Forward, she has 2 berths in the bow, an enclosed head with a shower in the head opposite an hanging locker. In the pilothouse is a raised dinette, helm station and galley. It would be tempting to rig an awning over the whole cockpit to give shade and shelter for open air living there.

Table of Particulars:

36' Power Dory	Imperial	Metric
Length Overall	36'-0"	10.97 m
Length Datum Waterline	31'-0"	9.45 m
Beam	11'-0"	3.35 m
Draft	3'-1"	0.94 m
Freeboard, Forward	5'-0¾"	1.54 m
Freeboard, Least	3'-0¼"	0.92 m
Freeboard, Aft	4'-2¾"	1.29 m
Displacement, Cruising Trim*	13,425	6,008 kg
Displacement-Length Ratio	201	
Prismatic Coefficient	.63	
Pounds Per Inch Immersion	983	
Auxiliary Horsepower	50	
Water, Gallons	50	189 litres
Fuel, Gallons	260	984 litres
Headroom	6'-0" - 6'-7"	1.83-2.01 m

*****Caution**: The displacement quoted here is for the boat in coastal cruising trim. That is, with the fuel and water tanks filled, the crew on board, as well as the crews' gear and stores in the lockers. This should not be confused with the "shipping weight" often quoted as "displacement" by some manufacturers. This should be taken into account when comparing figures and ratios between this and other designs.

Building *Badger*

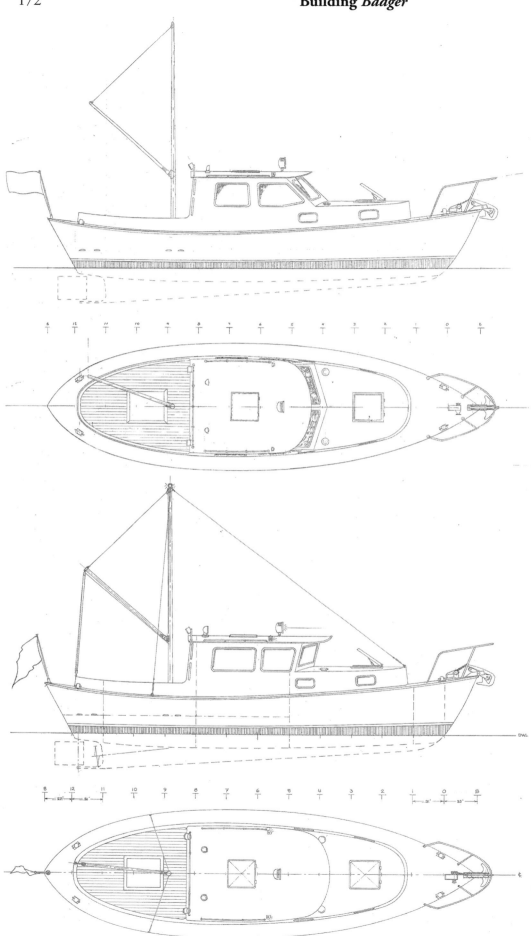

Building *Badger*

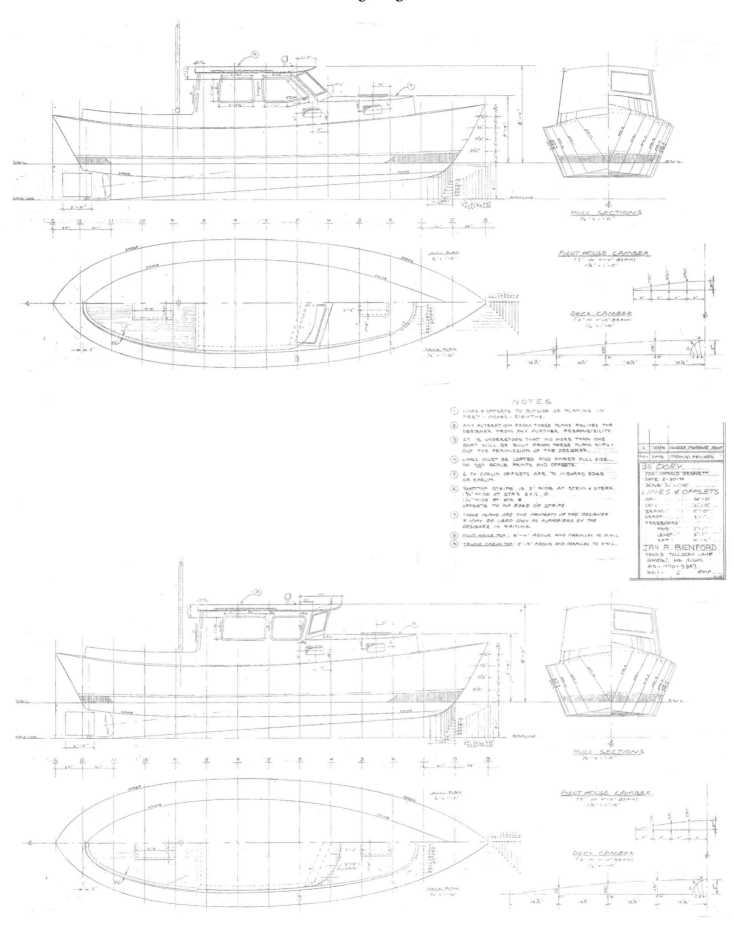

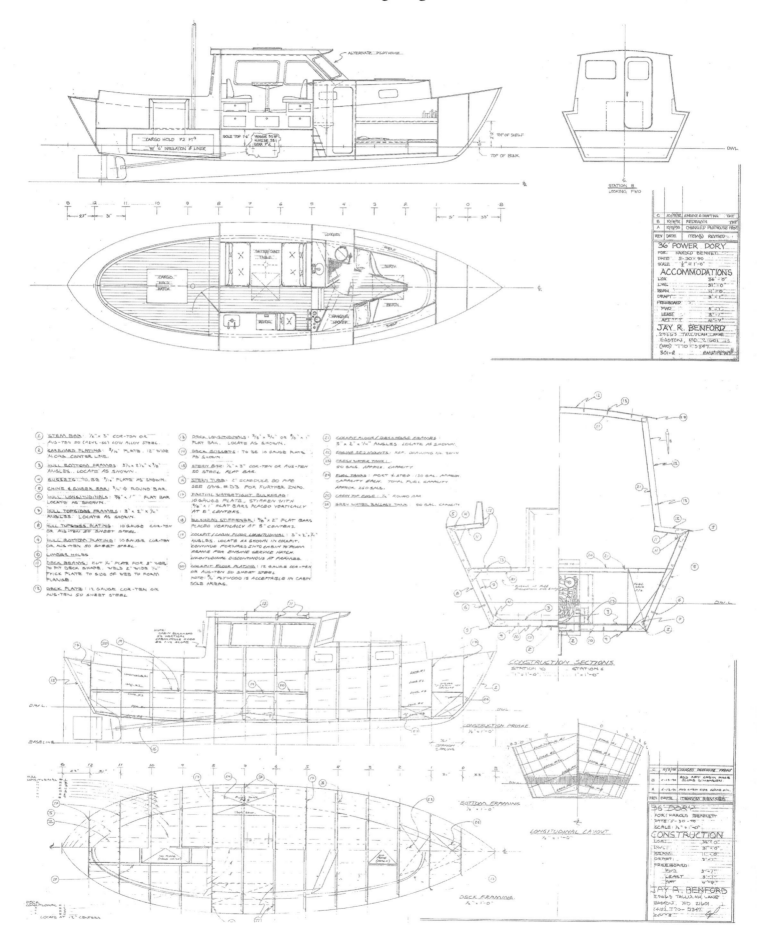

This is the preliminary sketch done to define the concept of the 36' Power Dory. You can see from the drawings how this turned out fairly close to the final version of the design.

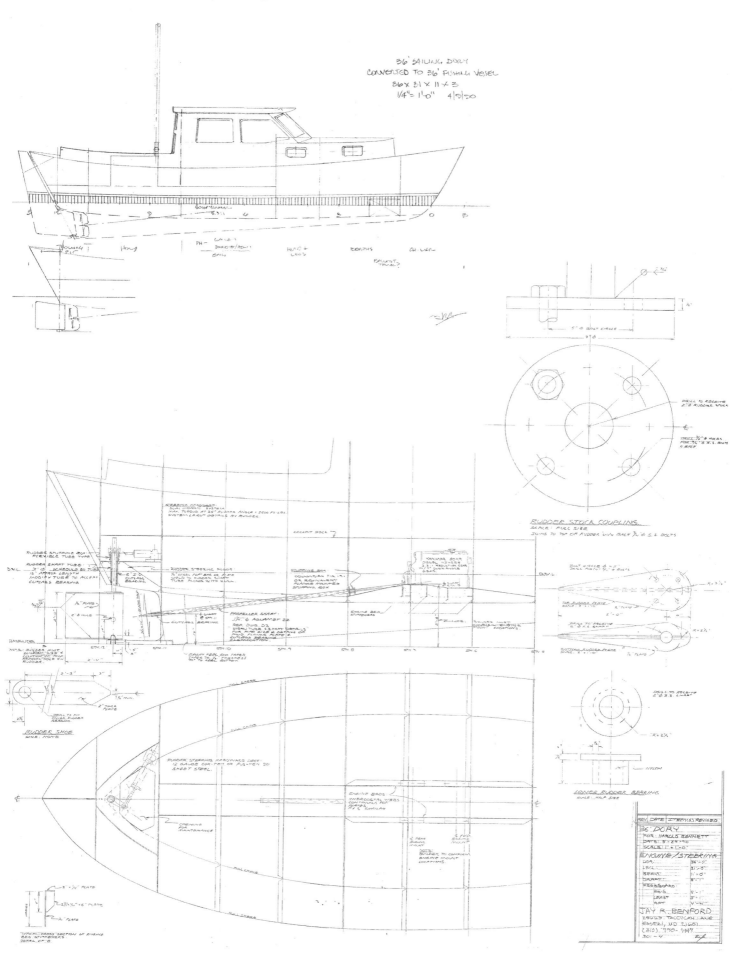

45' Sailing Dory
Design Number 323
1993

Table of Particulars:

45' Sailing Dory	Imperial	Metric
Length Overall	45'-0"	13.72 m
Length Datum Waterline	39'-1"	11.91 m
Beam	16'-0"	4.88 m
Draft—long, shallow keel (opt.)	4'-0"	1.22 m
Freeboard, Forward	6'-6"	1.98 m
Freeboard, Least	4'-3"	1.30 m
Freeboard, Aft	5'-3"	1.60 m
Displacement, Cruising Trim*	31,800 lbs	14,424 kg
Displacement-Length Ratio	238	
Ballast	10,000 lbs	4,536 kg
Ballast Ratio	31%	
Sail Area, Square Feet	1,209	112.32 m²
Sail Area-Displacement Ratio	19.27	
Prismatic Coefficient	.617	
Pounds Per Inch Immersion	1,603	
Auxiliary Horsepower	60	
Water, Gallons	300	litres
Fuel, Gallons	150	litres
Headroom	6'-6"	1.98 m

***Caution**: The displacement quoted here is for the boat in coastal cruising trim. That is, with the fuel and water tanks filled, the crew on board, as well as the crews' gear and stores in the lockers. This should not be confused with the "shipping weight" often quoted as "displacement" by some manufacturers. This should be taken into account when comparing figures and ratios between this and other designs. Also, loading down for voyaging and living aboard, as with the Hill's *Badger*, will add considerably to these figures, perhaps as much as 50%. Designs like the 34' *Badger*, which load down gracefully and still sail well, make a good choice for anyone wanting to go voyaging on a small income.

The 45' Sailing Dory was created for a fellow who wanted to run it as a charter operation on the Chesapeake Bay, taking two couples on outings. As the drawings show, this is an uncompleted design, shown here just because it's an interesting idea.

There are three pages of drawings shown here. The first is a couple of the preliminary ideas we looked at, trying to sort out an agreeable layout to separate the crew from the charter party. The second one shows the final version with two charter cabins with their own heads and showers, plus another space in the aft cabin for the crew. The pilothouse has the galley and social areas. We looked at several different rigs, and ended up with this one, thinking it would be functional and be an attention getter for the charter business. The third page has the scantling section, notes, and some more sections.

The side and forward decks are below the sheer line, giving secure feeling bulwarks. This generous on deck space will make for lots of opportunities for people to be on deck and feel comfortable about it.

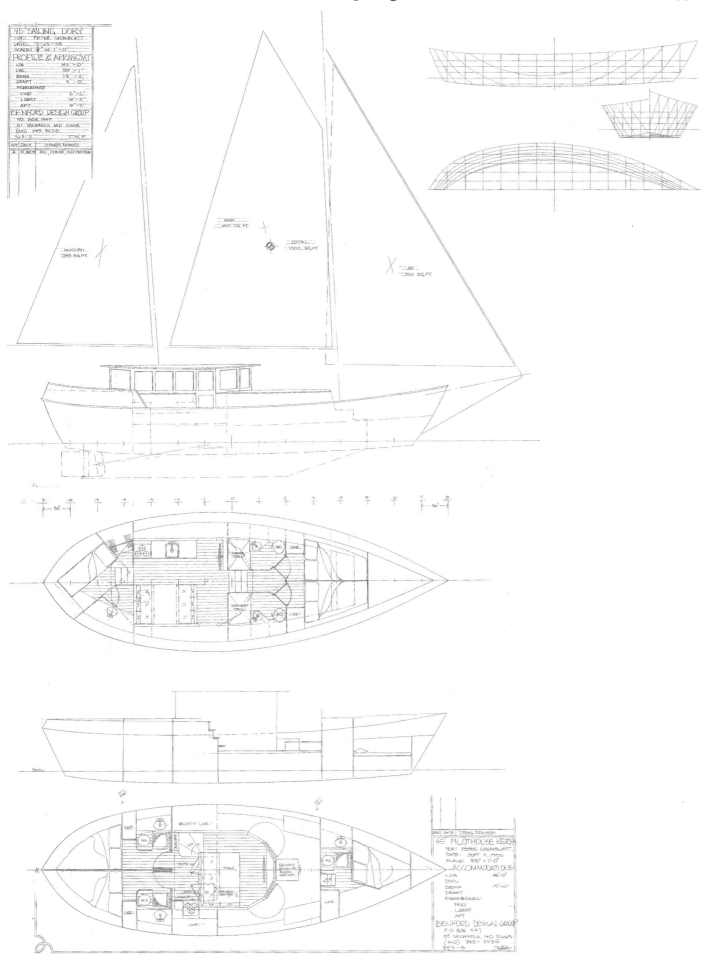

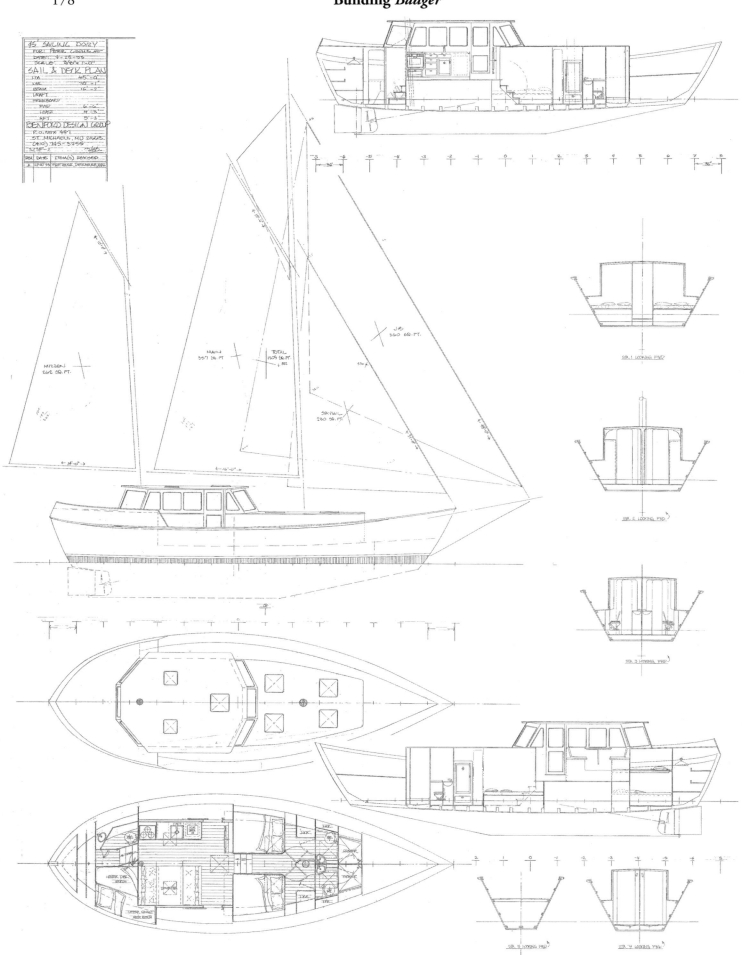

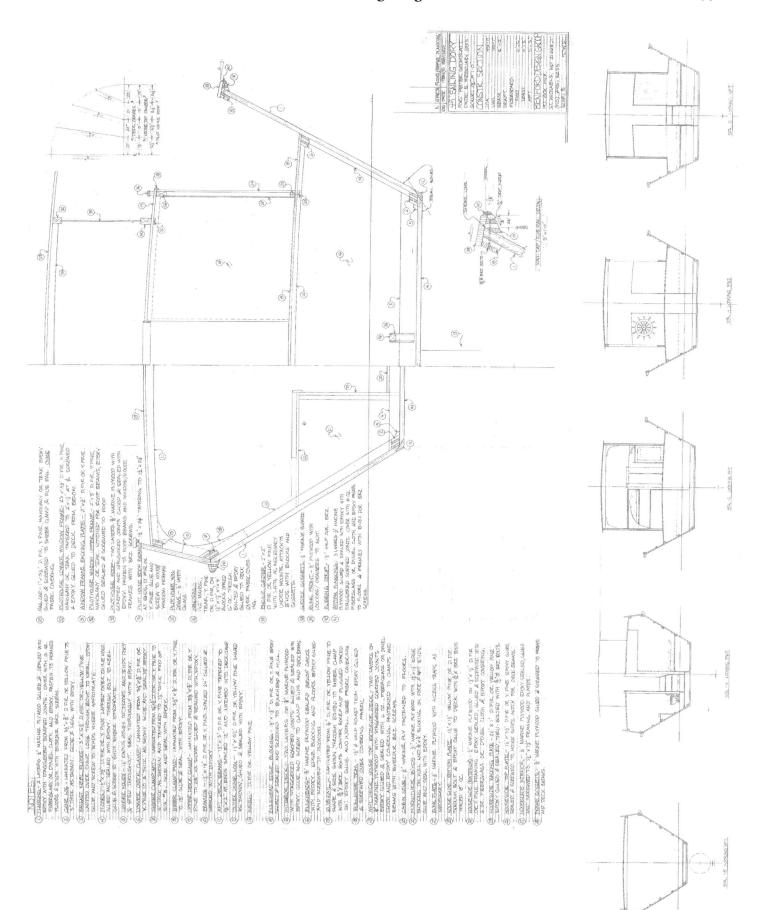

30'-5" & 31'-8" Sailing Dory
Design Number 370
2002

There are two versions of this design, a transom stern 30'-5" and a double ended 31'-8". They share a lot of drawings for the structure and rigs are identical, except in way of the stern. The basis for the boat was our design number 32, originally done in 1967, and the basis for several variations over the next couple decades.

The raised deck versions, designed with both a tombstone transom (30'-5") and the double-ender (31'-8") have junk rigs, single or double masted, available. While they might be thought

of as a "Baby *Badger*" they are capable boats in their own rights, whilst not having quite as much carrying capacity as the 34-footer does. Still, they would make good voyaging and liveaboard yachts.

Annie Hill, in **Voyaging On A Small Income**, recommended having a small transom for ease of mounting self-steering gear. While this might facilitate mounting the steering vane, there are other types that can readily mount on a double ender. The double ender will load down more gracefully and still not be dragging a transom through the water when heavily laden.

The indicated displacement is in coastal cruising use. For voyaging, I would assume that she would get loaded down a fair bit, like *Badger*, with the stores and supplies needed for such service. Fortunately, these dories take this loading gracefully, not unlike their predecessors the working dories which set out light and returned carrying tons of fish. At one time, when visiting aboard the Hill's 34' sailing dory, *Badger*, I noticed that she was floating eight to nine inches below where she was designed—with all their worldly possessions aboard. She still had good stability and was still sailing as a double ender.

There are a schooner and a single sail sloop version available. The interior on the sloop shows a sit-down chart table instead of the stand up table. This works because there is not a clearance problem with the main mast of the schooner version.

If you're not going long distance cruising and don't need the additional stores carrying ability of the larger boat, these are a good choice. The smaller size will be less expensive to build and outfit, pretty much in direct proportion to the displacement, 7,400 versus 10,400 pounds for the 34' *Badger*.

These simple-to-construct, no-nonsense sailing dories will perform well. Though flat-bottomed in dory fashion, when they're heeled over, they present a "v" to the water, and will move along at quite a good clip. They have 2,800 pound lead ballast on their keel, with plywood, hard chine construction over sawn frames. They have an enclosed head, functional galley, roomy area for dining and lounging and a double berth for a couple and settee berths for sleeping on a passage.

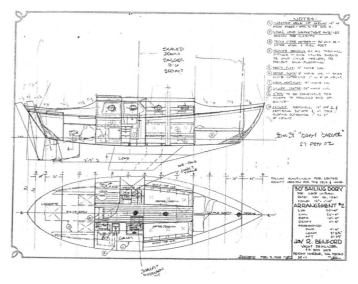

Preliminary Sketch done 27 Feb 02

Table of Particulars:

Baby *Badger* Sailing Dory	30'-5"	Metric	31'-8"	Metric
Length Overall	30'-5"	9.27 m	31'-8"	9.65 m
Length Datum Waterline	26'-0"	7.92 m	26'-0"	7.92 m
Beam	10'-0"	3.05 m	10'-0"	3.05 m
Draft	4'-0"	1.22 m	4'-3"	1.30 m
Draft—long, shallow keel	3'-3"	0.99 m	3'-6"	1.07 m
Freeboard: Forward	4'-6"	1.37 m	4'-3"	1.37 m
Freeboard: Least	2'-8¾"	0.83 m	2'-5¾"	0.83 m
Freeboard, raised deck	5'-2"		4'-11"	
Freeboard: Aft	3'-3½"	1.00 m	3'-0½"	0.93 m
Displacement, Cruising Trim*	7,400	3,357 kg	9,500	4,309 kg
Displacement-Length Ratio	188		241	
Ballast	2,680	1,216 kg	2,680	1,216 kg
Ballast Ratio	38%		30%	
Sail Area, Square Feet	500	46.45 sq m	500	46.45 sq m
Sail Area-Displacement Ratio	21.07		17.84	
Prismatic Coefficient	.58		.58	
Pounds Per Inch Immersion	685		685	
Auxiliary Horsepower	9		9	
Water, Gallons	25	95 liters	25	95 liters
Fuel, Gallons	25	95 liters	25	95 liters
Headroom	6'-1"	1.85 m	6'-1"	1.85 m

*__Caution__: The displacements quoted here are for the boat in coastal cruising trim (7,400 pounds) and average load for voyaging (9,500 pounds). That is, with the fuel and water tanks filled, the crew on board, as well as the crews' gear and stores in the lockers. This should not be confused with the "shipping weight" often quoted as "displacement" by some manufacturers. This should be taken into account when comparing figures and ratios between this and other designs.

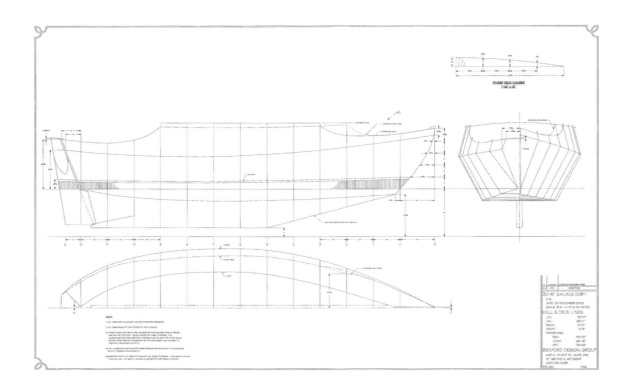

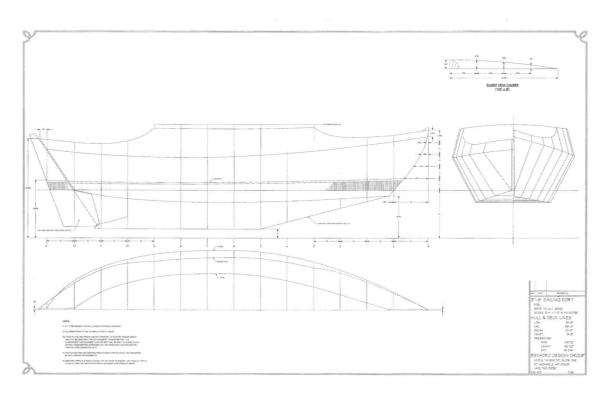

Building *Badger*

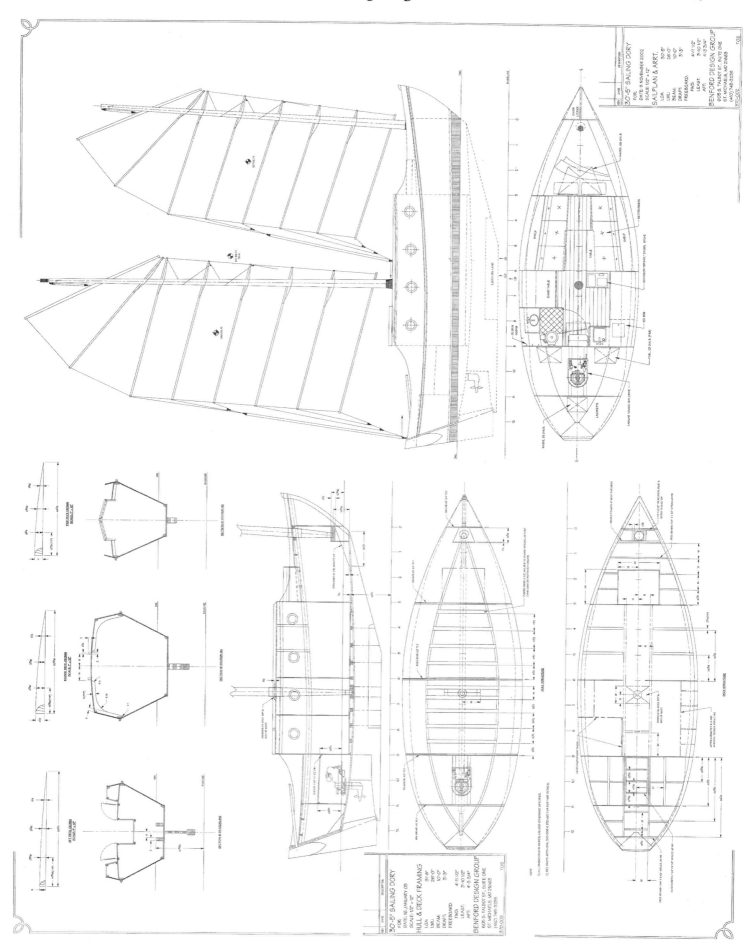

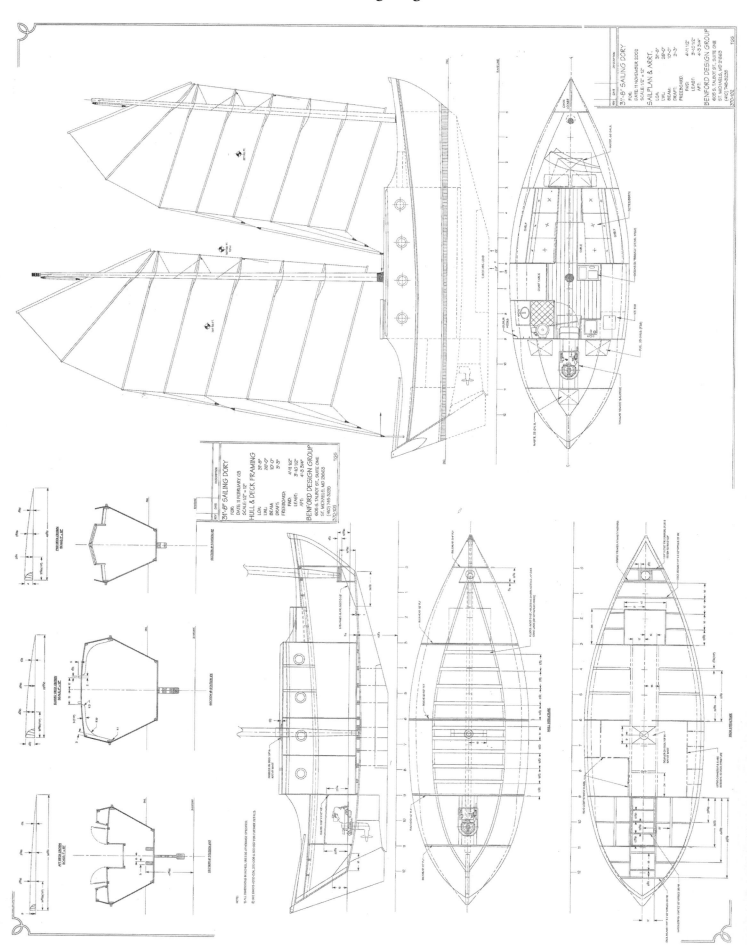

Building *Badger*

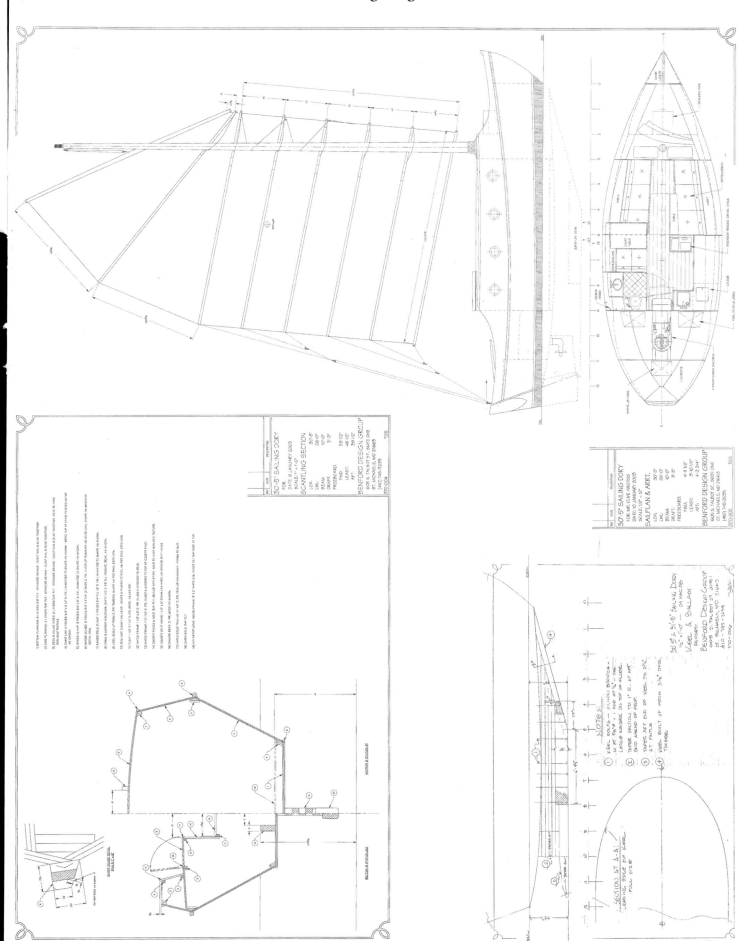

Building *Badger*

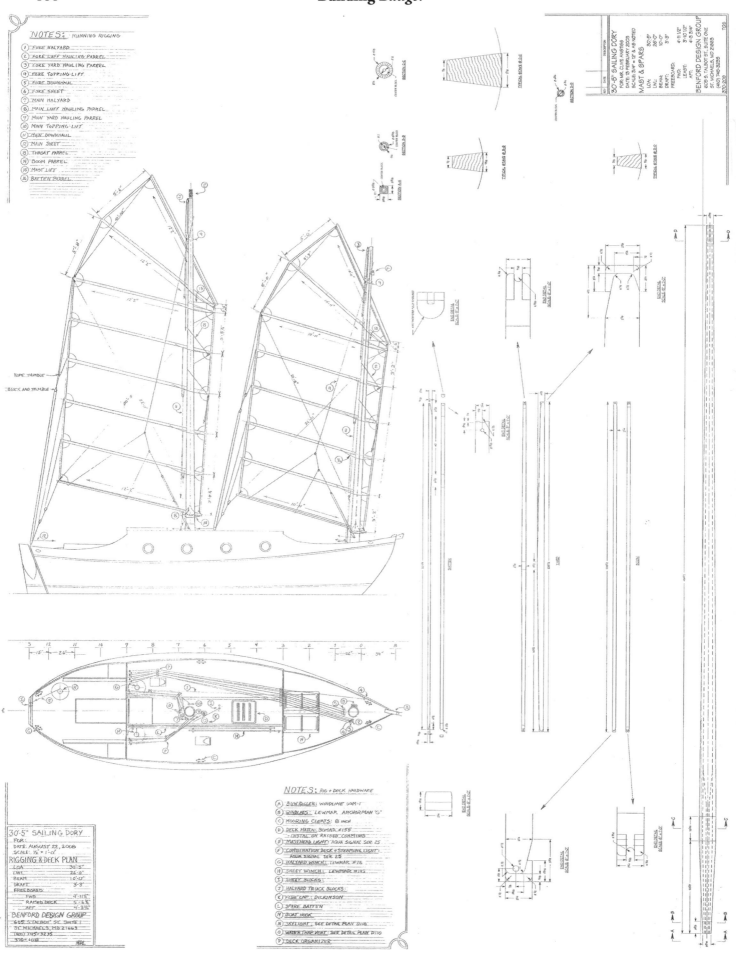

45' Sailing Dory
Design Number 384
2007

This boat has some interesting features requested by our client. Her rig is not too tall, but will provide good power for her. He liked the look of the 34 and 37½ and wanted a larger boat with that look and junk schooner rig. Her keel is a hollow welded steel fin, with the generator and electric motor in it, plus tankage for fuel and ballast. Some of the details on this are shown on the drawings on the following pages. Like ***Badger***, her galley and head are right alongside the companionway ladder, with chart table and dining area forward of them.

The accommodations has staterooms forward and aft. The after stateroom is the master, with a walk-around, centerline double berth. Outboard there are bookshelves for a considerable library, and over the berth is a large skylight for plenty of natural light and fresh air. The forward stateroom's single berths could be modified if there was another use in mind. In the bow is a large forepeak locker, with room for quite a lot of ship's gear. Hatches in the sole provide access to the generator and engine.

Table of Particulars:

45' Sailing Dory	Imperial	Metric
Length Overall	45'-0"	13.72 m
Length Datum Waterline	39'-9"	12.12 m
Beam	12'-0"	3.66 m
Draft—long, shallow keel (opt.)	4'-6"	1.37 m
Freeboard, Forward	5'-10"	1.78 m
Freeboard, Amidships	5'-8"	1.73 m
Freeboard, Aft	4'-11"	1.50 m
Displacement, Cruising Trim*	25,695 lbs	11,655 kg
Displacement-Length Ratio	183	
Ballast	5,000 lbs	2,268 kg
Ballast Ratio	20%	
Sail Area, Square Feet	1,010	93.83 m²
Sail Area-Displacement Ratio	18.56	
Prismatic Coefficient	.641	
Pounds Per Inch Immersion	1,403	
Auxiliary Horsepower	40	
Water, Gallons	200	757 litres
Fuel, Gallons	222	840 litres
Headroom	6'-3"	1.91 m

*Caution: The displacement quoted here is for the boat in coastal cruising trim. That is, with the fuel and water tanks filled, the crew on board, as well as the crews' gear and stores in the lockers. This should not be confused with the "shipping weight" often quoted as "displacement" by some manufacturers. This should be taken into account when comparing figures and ratios between this and other designs. Also, loading down for voyaging and living aboard, as with the Hill's ***Badger***, will add considerably to these figures, perhaps as much as 50%. Designs like the 34' ***Badger***, which load down gracefully and still sail well, make a good choice for anyone wanting to go voyaging on a small income.

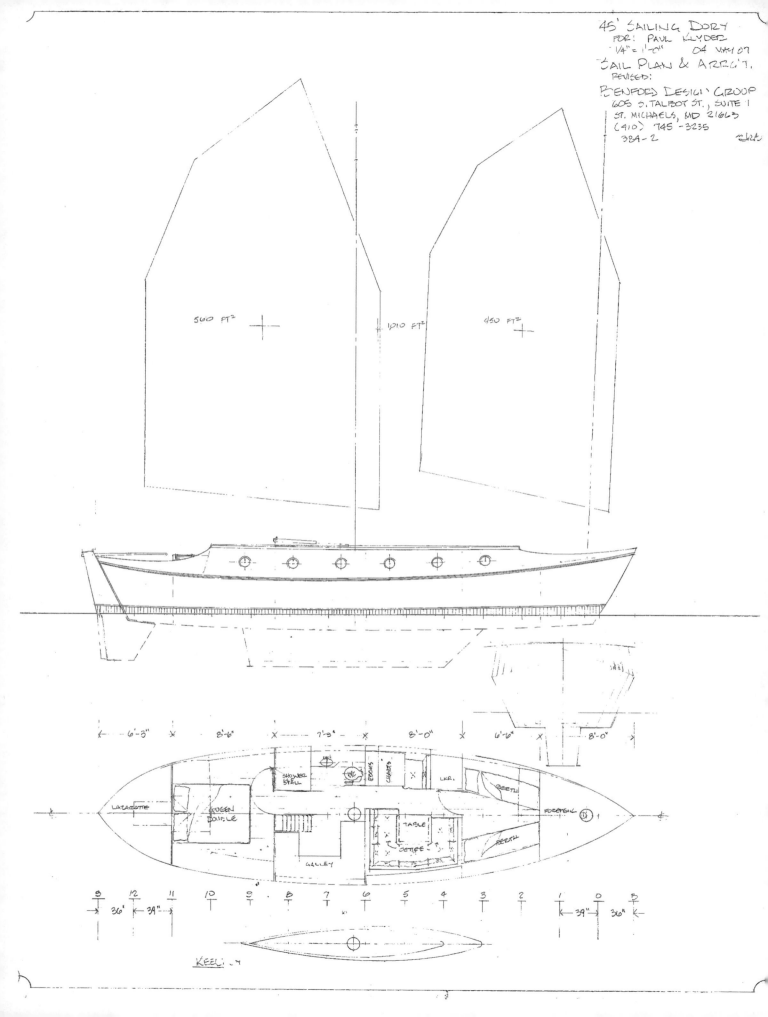

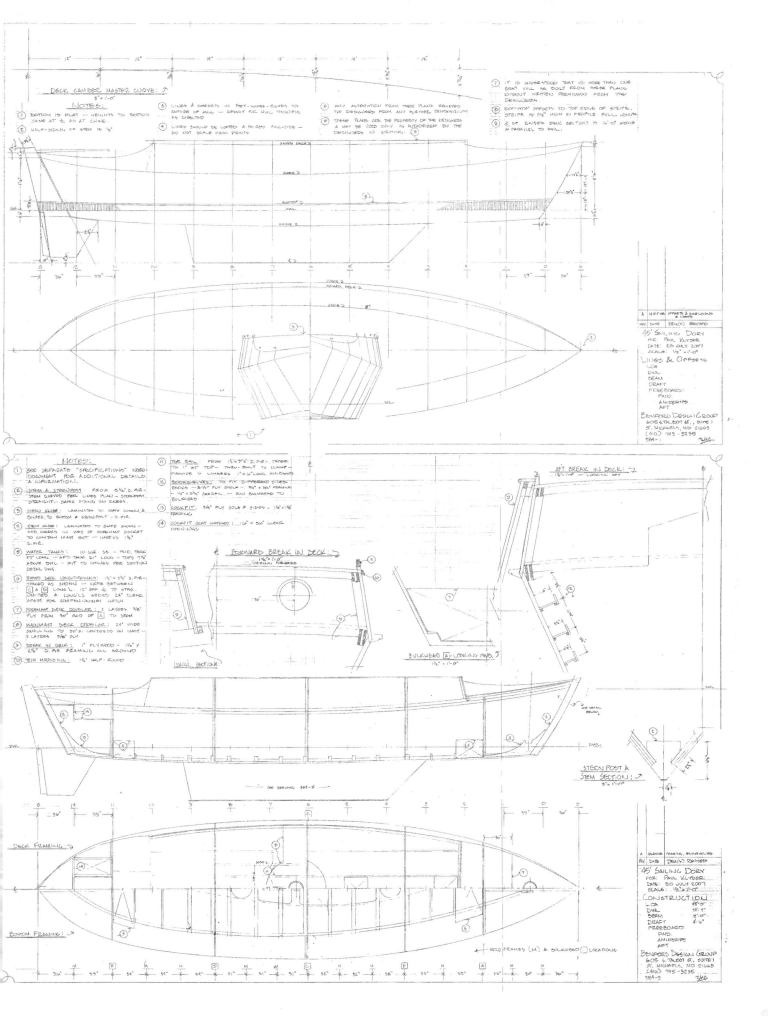

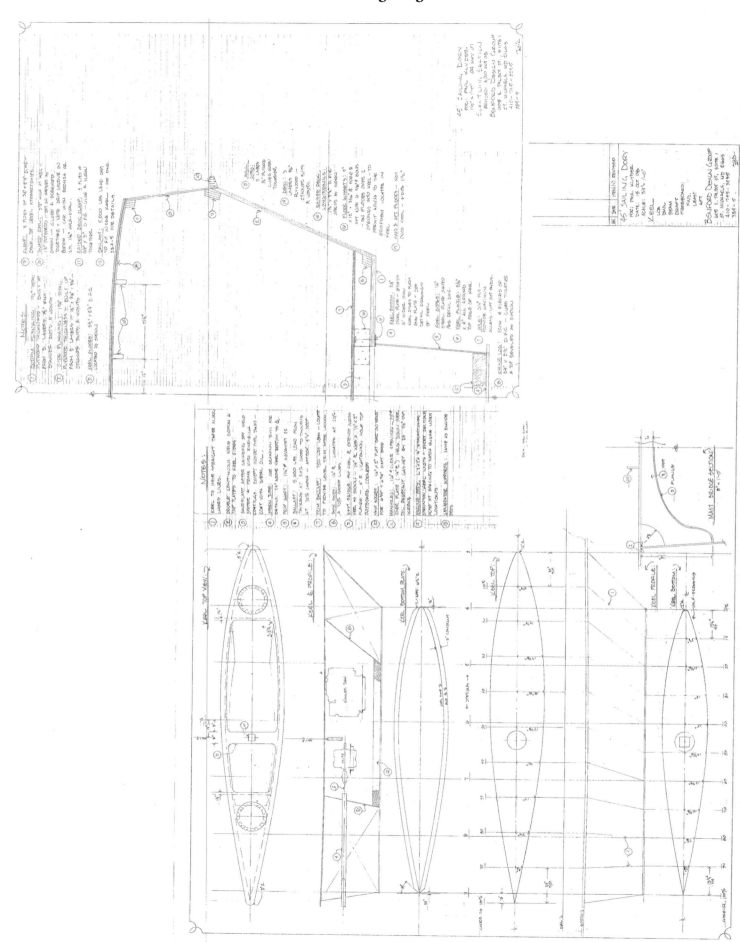

Glossary of terms found on drawings:

Aft: Towards the back (after) end or stern of the boat.
Athwartships: In a direction 90-degrees from the centerline of the boat, that is towards the side of the boat.
Barge: A rectangular shaped vessel, usually having sloped panels on the bottom forward and aft.
Beam: The greatest width of the boat. Usually taken to the structure of the boat, not over the guardrails.
Bow: The front end of the boat.
Bulwark: A solid panel, extending upwards from the deck, providing shelter and security for the people on deck. Often it is an extension of the hull sides, usually with tumblehome.
Chine: The angular joint, joining two hull panels, where the bottom and sides meet. That is, the ribs of the boat have an angular joint in them.
Coaster: A vessel that usually cruises along the coast. This can and does involve open water passages. Not intended to imply a lack of seagoing ability or capability.
Displacement, cruising trim: The weight of the boat with its usual load of people and supplies aboard. The boat displaces it's own weight in water—the volume of water displaced equals the boat's weight.
Displacement-length ratio: Non-dimensional ratio indicating whether boat is "light" (under 250) and possibly driven quite fast or "heavy" (over 350) and better suited to slower speeds. "Heavy" is likely to have more comfortable motion underway and at anchor, with its weight damping the motion.
Double-ended: A boat with no transom at the stern, but rather having the sides coming back together at the centerline.
Draft: Distance the deepest part of the boat is immersed in the water.
D/W: Dishwasher
DWL: Designed or datum water line.
English: See Imperial.
Fantail: An overhanging counter stern, usually elliptical in plan view.
Flare: The outward slope of the hull sides, most often found at the bow.
Foc'sle: A contraction of forecastle, the forward or bow cabin.
Fore and aft: In the direction parallel to the centerline of the boat, towards where the boat is headed or where it has been.

Forward: Towards the front end or bow of the boat.
Freeboard: The distance from the water's surface to the deck (sometimes measured to the top of the hull or top of the bulwark.)
Fuel tankage: Total capacity of the fuel tanks aboard.
Galley: The "kitchen" on the boat.
Guard Rails: Outward structural projections to provide protection for the planking, usually heavily built.
Head: The "bathroom" on the boat. Sometimes used to specifically note the water closet or toilet.
Headroom: Clear vertical distance inside the boat for people to move about inside the boat without bumping their heads.
Imperial: Measurements taken in units of feet, inches, pounds and gallons. Also known as the English system of measurement.
Keel: The longitudinal member of the structure providing strength along the centerline of the vessel.
Length overall: Total length of the structure. Does not usually include portions of the boat that hang beyond the main part of the structure, such as bowsprits.
Length designed waterline: Total length of the boat measured where it meets the surface of the water.
Metric: Measurements taken in units of meters, kilograms and liters.
Packet: Traditionally, a vessel carrying freight, usually on a set route and schedule.
Particulars: A table of measurements.
Pilothouse: The space aboard devoted to the controls for operating the boat.
Planking: The outer skin of the boat. Also called plating on a metal boat.
Plating: See Planking.
Port: The left hand side of the boat when facing forward. Also, an opening in the side of the boat.
Portlight: An opening port (window).
Prismatic coefficient: The ratio of the area of a prism, with a section equal to the largest cross-section of the boat, to the total volume of the boat. The lower the number the more easily driven at lower speeds. A higher number, over 0.7, is usually associated with boats intended for higher speeds.
P/S: Port and Starboard.

Rabbet: The joint where the planking or plating joins the keel.
Raised Deck: A style of design in which the deck is raised above the sheer line to create a more voluminous interior.
Ribs: Transverse, or athwartships, structural framing members.
Round bilged: The hull shape where the ribs are smoothly curved, with no breaks in them, from keel to sheer.
Rudder: The movable blade, located directly aft of the propeller, that redirects the water flow and causes a reaction that steers the boat.
Saloon: Main living space aboard, like a living room ashore. Large ships sometimes also have dining saloons, where people congregate to be fed. Contrast with **salon**, a place to have one's hair styled.
Sheer: The top edge of the hull at the side. Usually it's the widest part of the boat, unless it has tumblehome, and sweeps from bow to stern.
Single Screw: A single propeller on centerline of the boat.
Spud: A vertically movable pole that drops through a tube built into the hull ("stick-in-the-mud") to hold the boat fixed in position, or to provide a point for a pivoting maneuver. Most useful in shallower waters or in areas with smaller tidal ranges.
Starboard: The right hand side of the vessel when facing forward.
Stem: The leading edge of the hull, the extension of the keel that sweeps up and rises to join the deck.
Stern: The aft or back end of the boat.
Tankage: The capacity of the tanks.
T/C: Trash compactor.
Transom: The panel at the stern, which joins the sides and bottom. It is often curved in plan view for strength and improved appearance.
Tumblehome: Inward slope of the boat sides, narrowing as it goes up.
Twin Screw: Two propellers (screws). The use of two usually well-spaced apart props gives good leverage when maneuvering in close quarters.
Water tankage: Total capacity of the water tanks aboard.
W/D: Washer and Dryer.
Yacht: A vessel of pleasure or state. May be of any size.

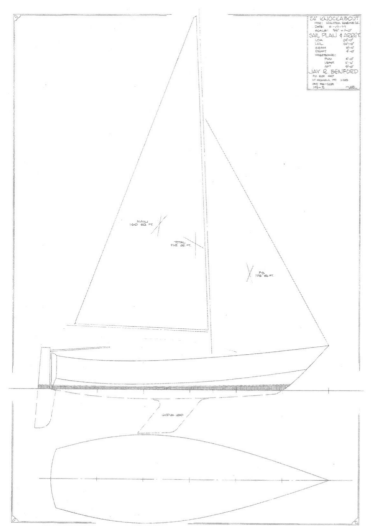

Other Dory Design Ideas—shown here, clockwise from top left, are a 24' double chine knockabout, a 20' double-ender, and a 24' little schooner. The possibilities are endless. What sort do you think would be fun?...

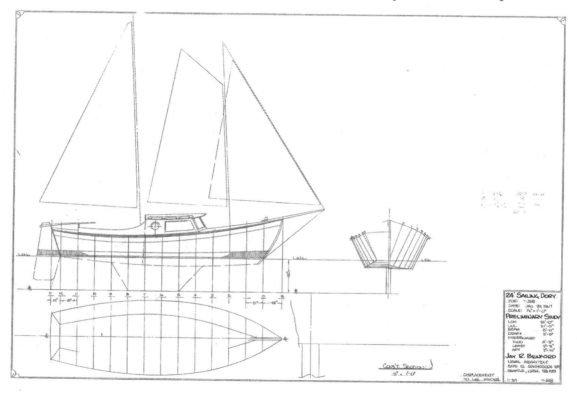